Heat and Mass Transfer

Series Editors

Dieter Mewes, Universität Hannover, Hannover, Germany
Franz Mayinger, München, Germany

This book series publishes monographs and professional books in all fields of heat and mass transfer, presenting the interrelationships between scientific foundations, experimental techniques, model-based analysis of results and their transfer to technological applications. The authors are all leading experts in their fields.

Heat and Mass Transfer addresses professionals and researchers, students and teachers alike. It aims to provide both basic knowledge and practical solutions, while also fostering discussion and drawing attention to the synergies that are essential to start new research projects.

More information about this series at http://www.springer.com/series/4247

Vladimir Danilov · Roman Gaydukov ·
Vadim Kretov

Mathematical Modeling of Emission in Small-Size Cathode

 Springer

Vladimir Danilov
National Research University
Higher School of Economics
Moscow, Russia

Roman Gaydukov
National Research University
Higher School of Economics
Moscow, Russia

Vadim Kretov
National Research University
Higher School of Economics
Moscow, Russia

ISSN 1860-4846　　　　　　　ISSN 1860-4854　(electronic)
Heat and Mass Transfer
ISBN 978-981-15-0197-5　　　ISBN 978-981-15-0195-1　(eBook)
https://doi.org/10.1007/978-981-15-0195-1

This Springer imprint is published by the registered company Springer Nature Singapore Pte Ltd.
The registered company address is: 152 Beach Road, #21-01/04 Gateway East, Singapore 189721, Singapore

Preface

The study described in this book began several years ago when Prof. M. V. Karasev acquainted the authors with two MIET (National Research University "Moscow Institute of Electronic Technology") researchers, N. A. Dyuzhev and M. A. Makhiboroda who reported at the seminar that the mathematical modeling of electron emission from small-size cathodes has not yet been investigated full enough. The main problem in this study is the modeling of heat transfer with possible cathode melting.

The global (this term is explained in the book below) description of the heat transfer with melting taken into account can be obtained using the phase-field system, i.e., in the framework of the contemporary generalization of Stefan-type problems.

The phase-field model has already been studied during approximately 15 years in the mathematical literature but, as it turned out, the specialists in the field of electronics knew nearly nothing about it. The result of this model adoption is discussed in the book proposed to the reader.

It should be noted that the electron tunneling process was studied (mainly, from the physical viewpoint) in detail for semiconductors and metals in G. Fursey's book.[1] But the numerical modeling described in that book gave a very high temperature of the cathode but without taking the possibility of melting into account.

The proposed monograph mainly deals with mathematical modeling, namely it describes the mathematical model of heat transfer in a silicon cathode of small (nano) dimensions with the possibility of partial melting taken into account. This is a mathematically very difficult problem, and the properties of the solution of the phase-field system underlying our study have not yet been described completely. We present only well-known analytic results, but the basic idea in this book is the following one.

Instead of a pure mathematical approach, we apply an approach based on the understanding of the solution structure and computer simulation. The understanding of the solution structure is based on the construction of asymptotic solutions and an

---

[1]G. Fursey "Field Emission in Vacuum Microelectronics", Springer (2005).

analysis of their properties. The further comparison of the results of numerical experiments with the asymptotic solutions allows us to determine the process behavior and its characteristics.

Further, note that not all details of physical processes in the cathode (especially in the case of appearance of the liquid phase) have been described in the literature. Therefore, in numerical experiments, we made several additional assumptions. Moreover, in this book, we briefly consider some information given in the physical literature which (from our viewpoint) explains the properties of the mathematical model.

All physical parameters contained in our model can be found in the physical literature, i.e., in different reference books. The only exception is the expression describing the Nottingham effect. All physical constants in this expression are known, but the formula itself is obtained on the basis of tunneling through the potential barrier from half-plane to space. Thus, the actual geometry of the cathode is not taken into account in this formula. Usually, in problems of such a type, the cathode geometry is taken into account by using the so-called form-factor. We follow this way choosing a multiplier in the expression for the heat flow from the emitter surface. This multiplier is the only "adjusting" parameter in our problem and can easily be determined experimentally by comparing the theoretical value of the emission flow density (from the plane surface of the emitter) with the actual emission current observed in experiments.

The book is organized as follows. In Chap. 1, we describe the history of discovering the electron emission phenomenon and its types. We also present the mathematical statement of the problem of the field emission from semiconductor cathodes of small dimensions.

In Chap. 2, we briefly discuss some information from the solid-state physics, in particular, formulas for specific conductivity, thermoelectric height, and the Thompson coefficient for semiconductors. We also mathematically describe the tunneling process through the potential barrier on the cathode–vacuum interface, write a formula for the field emission current density in the case of metals, and describe its specific features in the case of field emission from semiconductor cathodes. At the end of the second chapter, we describe the Fowler–Nordheim theory and the Nottingham effect in the case of field emission from metals.

In Chap. 3, we discuss the phase-field model and the basic properties of its solutions. We also present formulas for asymptotic solutions of the phase-field system in the simplest case and in some special cases.

In Chap. 4, we present formulas for the numerical solution of the system of phase-field equations and an algorithm for solving the problem numerically (including its realization by contemporary multiprocessor systems and hybrid systems based on graphic accelerators) and discuss the results of numerical experiments.

Preface to the English Edition

In addition to the Preface to the Russian edition, we would like to make some remarks. The book deals with the physical problem—electron emission from a small-size cathode.

We initially intended to solve the problem of phase transition concerning the field emission of electrons from a small-size cathode. Obviously, the small size of the cathode and a relatively strong electric current density through it could result in melting of the cathode core. Thus, it was necessary to solve the Stefan problem taking into account the large curvature of the boundary between phases (the latter follows from the small size of the cathode).

If one begins to think about numerical calculations of the Stefan problem solution, then he immediately comes to the idea of adaptive mesh or something like this. A significant difference between geometric sizes (extremely small near the top of the cathode and relatively large near the bottom) results in additional troubles. Thus, to obtain the global solution of the phase-field system approximating the solution of the original problem seemed to be the most simple way to solve the problem.

We would like to stress that for the research work described in the book, we have to use relatively contemporary mathematical tools, especially the last results of the theory of generalized functions.

The most important point here is the construction of the definition of generalized solution to the phase-field system; see Sects. 3.4 and 4.6. As was discovered, the initial stage of melting required a special construction of liquid-phase nucleus for its description, where we actively used the concepts of mushy region and interaction of nonlinear waves; see Sect. 3.5.

And finally, we propose to use visualizations simultaneously with pure mathematical considerations to investigate the mathematical model; see Sect. 3.5.

The mathematical background is not the main part of this book, but, by our opinion, it would be better to learn about it before one starts his own investigations in the field described here.

For this edition, we checked and corrected many misprints and added some comments through out the text.

Moscow, Russia   Vladimir Danilov
July 2019   Roman Gaydukov
                                                                                    Vadim Kretov

Acknowledgements

The authors express their deep gratitude to Professor of Department of Applied Mathematics HSE University (National Research University Higher School of Economics), M. V. Karasev, for the useful discussions and comments when the manuscript was prepared in the Russian language.

The authors thank Rector of MTUCI A. S. Adzhemov (Moscow Technical University of Communication and Informatics) for the help in publishing the monograph (in Russian).

The authors also thank Professor of Department of Applied Mathematics HSE University, V. M. Chetverikov, for the useful remarks.

The study of V. G. Danilov, R. K. Gaydukov, and V. I. Kretov was implemented in the framework of the Basic Research Program at the National Research University Higher School of Economics (HSE University).

Contents

Chapter 1
Introduction

Abstract In this chapter, a general description of the mathematical model of heat transfer and field emission is presented.

1.1 Brief History of the Electron Emission Discovery

The emission of charged particles, especially of electrons, from a solid or a liquid into a vacuum is a very interesting phenomenon which finds practical applications in various fields. It suffices to note that this phenomenon underlies the operation of electronic beam devices, electron microscopes, photocells, and so on.

The emission phenomenon has been known since the 1850s. In 1839, Alexandre-Edmond Becquerel observed [1] the photoelectric effect in elecrolyte, see Sect. 1.2. Further, in 1873, Willoughby Smith discovered that selenium is a photoconductive element[1] [39]. Later, in 1887, this effect was studied by Hertz [23]. When working with an open resonator, he noticed that the ultraviolet light significantly simplifies the spark advancing in a zinc discharger.

In 1888–1890, the photoelectric effect was systematically studied by the Russian physicist Stoletov [41–43]. He made several important discoveries in this field, including the first law of the external photoelectric effect (photoemission effect). But the photoelectric effect was explained only in 1905 by Albert Einstein (for this he received the Nobel Prize in 1921) based on the Max Planck hypothesis of the quantum nature of light. The studies of the photoelectric effect were some of the first quantum mechanical investigations.

The first report on electron emission of a different type, i.e., the thermal electron emission (see Sect. 1.2), was made by Frederick Guthrie in 1873 [22]. When observing charged bodies, he discovered that a negatively charged red-hot iron sphere somehow discharges in the air thus losing the charge. He also noted that there is no discharge if the sphere is charged positively. The first investigations of thermal electron emission by Hittorf [24–28], Goldstein [20], Elster and Geitel [12–16] should

---

[1]The photoconductivity is the effect of an increase in the electric conductivity of a semiconductor under the action of electromagnetic radiation.

© Springer Nature Singapore Pte Ltd. 2020

V. Danilov et al., *Mathematical Modeling of Emission in Small-Size Cathode*, Heat and Mass Transfer, https://doi.org/10.1007/978-981-15-0195-1_1

also be mentioned. This effect was reopened by Thomas Edison in 1880 when he tried to understand the cause of the wire filament fracture in the incandescent lamp.

Since the word "electron" had not been known as a separate physical particle until 1897, when Thomson's work was published, the word "electron" had not been used in the discussions of experiments until this publication. After the electron was discovered, the prominent English physicist Owen Willans Richardson started to investigate a phenomenon which he later called the "thermionic emission", and in 1928 he received the Nobel Prize in physics "for his work on the thermionic phenomenon and especially for the discovery of the law named after him" [35].

In the second half of the 19th century, the phenomenon of field emission (see Sect. 1.2) was also discovered. This phenomenon was first revealed by R. W. Wood in 1897. The first attempt of theoretical justification of this process was made by W. Schottky in 1923 [36]. His hypothesis was that the electrons emit through a surface potential barrier that decreases because of the applied electric field. In 1928, R. H. Fowler and L. W. Nordheim first proposed a theoretical explanation of the field emission effect. They were the first who obtained the dependence of the field emission current density on the electric field strength [18].

One of the latest emission phenomena, i.e., the explosive electron emission (see Sect. 1.2), was discovered in 1966 by S. P. Bugaev, P. N. Vorontsov-Vel'yaminov, A. M. Iskol'dskii, G. A. Mesyats, D. I. Proskurovskii, and G. N. Fursey, a group of scientists at Tomsk Institute of Automated Control Systems and Radioelectronics, Institute of Automatics and Electrometry and Institute of Atmospheric Optics, Siberian Branch of the USSR Academy of Sciences, and Leningrad State University.

1.2 Types of Electron Emission

At present, several types of electron emission are known: thermionic emission, photoemission, secondary electron emission, field emission, and explosive electron emission.

The *thermionic emission* is the phenomenon of electron emission by heated bodies. The thermionic emission is widely used in vacuum and gas-filled devices such as electron tube, cathode-ray tube, etc.

The *photoemission* or the *external photoelectric effect* is the emission of electrons from a matter under the action of electromagnetic radiation incident on its surface. This phenomenon is used in numerous devices which contain vacuum and gas-filled photocells, photoresistors, solar batteries.

The *secondary electron emission* is the electron emission from the solid surface bombarded by a beam of electrons. This phenomenon is used in photoelectron multipliers to amplify weak photoelectric currents, in electron-beam lithography, and so on.

The *field emission* is the electron emission by conducting solid and liquid bodies under the action of an external electric field without preliminary excitation of these electrons, i.e., without any additional energy expenditure, which is typical of the other types of electron emission. This phenomenon occurs because of the electron tunneling through the potential barrier near the body surface. This tunneling becomes possible because of the bending of the potential barrier under the action of a strong external field applied to the emitter surface. This type of emission is also called the *cold emission*.

The *explosive electron emission* is the electron emission under the action of local explosions on microscopic domains of the emitter. The explosive emission permits obtaining the maximal current density of all types of electron emission, which finds application in pulse generators of powerful electron beams and gas laser pumping. This emission was also used to design high-current vacuum diodes [3].

In this monograph, the field emission at large temperatures (also called the *thermo-field* emission, see Sect. 2.4) is studied in detail for small-size conic cathodes manufactured of semiconductors (more precisely, the silicon cathodes are considered). Such cathodes are shown in Figs. 1.1 and 1.2.

Cathodes similar to those considered in this book find application in field emission microscopy [31] and microscopy of different types and in electron-beam lithography. Research on constructing a display based on field emission cathodes of small dimensions was also carried out, but the serial production of such displays is currently an open problem [4, 38]. A more detailed description of the field emission can be found in monographs [11, 19]. In [11], some detailed information about the technology of manufacturing the emitters and various designs of field emission cathodes is given, the results of studies and application of new nanomaterials and carbon nanotubes

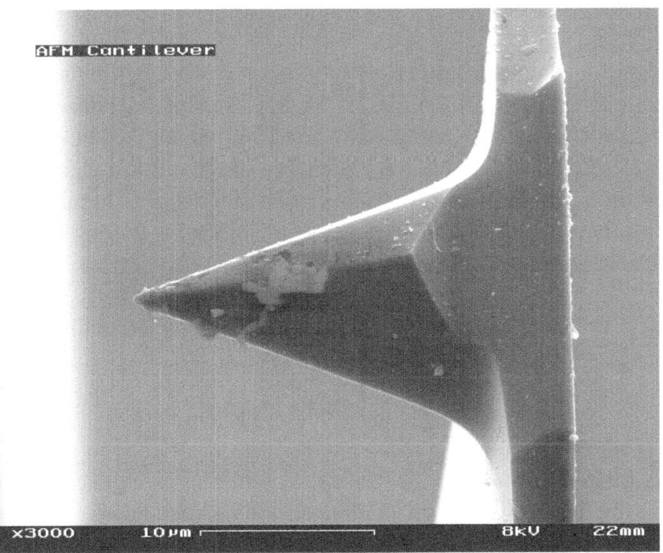

Fig. 1.1 AFM (used) cantilever in Scanning Electron Microscope, magnification 3000x (*Source* © SecretDisc ON WIKIPEDIA—CC-BY-SA 3.0, https://commons.wikimedia.org/wiki/File:AFM_ (used)_cantilever_in_Scanning_Electron_Microscope,_magnification_3000x.GIF)

Fig. 1.2 Cathode studied in [10]

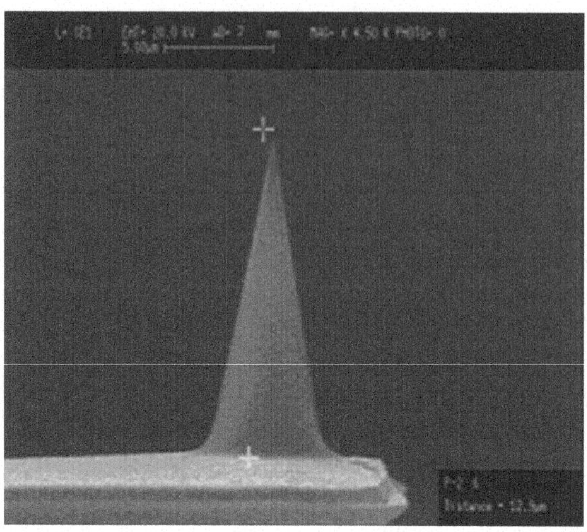

for designing field emission cathodes are discussed, and the applications of field emission cathodes in new light sources and flat-panel displays, microwave devices and X-ray tubes are considered.

1.3 Statement of the Problem

The problem of mathematical description of thermal processes in the field emission cathodes has a long history [5, 17, 30, 32, 34, 44]. A nontrivial phenomenon accompanying this process is the Nottingham effect [33], which is the cooling (heating) of the cathode provided that the average energy of the emitted electrons is greater (lower) than the Fermi level energy (see Sect. 2.4.7 for detail). This effect arises from the difference between the average energy of the electrons leaving the cathode and the average energy of the electrons approaching the emitter surface from the electric circuit.

This effect is mathematically described by the nonlinear condition of the third kind

$$\left.\frac{\partial T}{\partial \mathbf{n}}\right|_{S_e} = f_{\mathrm{Nott}}(T, j_{\mathrm{em}}, E_{\mathrm{F}}), \tag{1.1}$$

where T is the temperature on the surface S_e emitting electrons, $\mathbf{n}$ is the normal on the emitting surface S_e which is directed into a vacuum, j_{em} and E_{F} are the emission current density value and the electric field strength value at the points S_e. The function $f_{\mathrm{Nott}}(T, j_{\mathrm{em}}, E_{\mathrm{F}})$ can be written explicitly (see (1.10) and Sects. 1.4 and 2.4 for details) and has the property that, for a certain value $T = T^*$ called the inversion temperature, this function changes the sign. For $T < T^*$, the Nottingham boundary condition is the heating condition, and for $T > T^*$, the cooling condition. The quantity T^* is calculated for different materials (there is an a analytic expression for T^* in terms of the parameters characterizing the cathode material, see (2.116), (2.121)).

The goal in this work is to analyze the mathematical model of the heat propagation process, including the melting, in the case of field emission from a silicon pointed cathode of small (nano) dimensions. The main problem is to understand under which conditions the cathode point solidification occurs due to the Nottingham effect despite a large value of the emission current. Such experiments are described, for example, in [10]. This strange (at the first sight) phenomenon has not yet been mathematically investigated in detail. We can show a picture illustrating this phenomenon, see Fig. 1.3. Unfortunately, the quality of this picture is not good (we could not find a better one), but one can see that the melted area does not include the cathode vertex. The picture was made after the cathode solidification, and one can see that the cathode vertex preserved its shape.

In the system under study, the following two specific features play an important role. First, silicon is a semiconductor, and hence the Thomson effect can be neglected (see Sect. 2.3, and [2, 21, 29, 37, 40] for more details). Second, because of a sufficiently high thermal conductivity of silicon (cathode material) and small dimensions of the cathode, the dimensionless thermal conductivity coefficient turns out to be large, which is intuitively clear, i.e., the temperature is levelled rather fast in a small domain. This fact introduces significant difficulties into the computational algorithm (the large thermal conductivity coefficient is equivalent to large times).

Another important consequence of small dimensions (small radius of the top of the cathode) is a very sharp increase in the electric field strength near the top of the cathode as compared to the strength of the applied external field.

Fig. 1.3 Conic cathode after the end of emission, from [10]. One can see that the melting area does not contain the top of the cathode

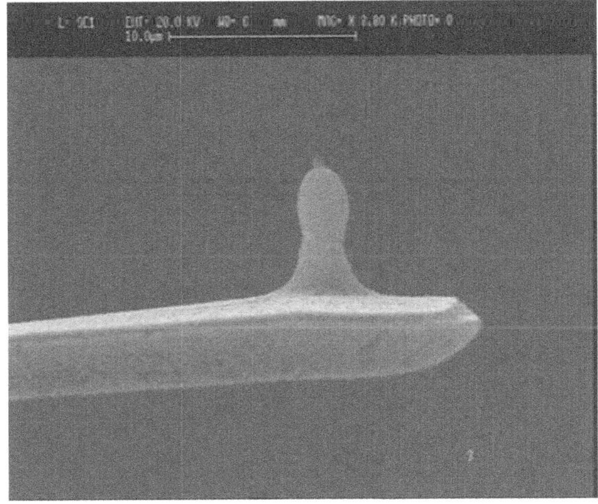

Significant additional difficulties arise because of the two-phase state of the heat conducting medium (solid and liquid) and the necessity to describe the motion of the heat front (the interface between the phases). Here we use the recently developed new mathematical approaches [6–9] to construct solutions of differential equations for multiphase media.

When modeling the field emission of electrons, we do not take into account mechanical stresses in the cathode material and do not consider the variations in the nanotcathode geometry.

We compare our results with the data of experimental studies [10], where the authors were inclined to assume that the process of melting of a certain layer of the cathode and its subsequent solidification was observed in experiments, but it was not clear at what particular time the solidification occurs, either after the current is switched off or already in the process of field emission. The results of our mathematical modeling with the parameters considered in [10] do not give the solidification pattern. But if the parameters are changed, then the effect of the cathode point solidification is obtained in our model precisely when the current flows through the cathode in the process of field emission. This effect is described in more detail and demonstrated in Chap. 4.

A detailed mathematical theory including the modeling of melting and crystallization is described in Chap. 3. Here we only note that the Nottingham effect is not monotone with respect to the temperature. As the temperature increases, the right-hand side of (1.1) also increases, which results in the cooling of the emitting surface. Because of a high thermal conductivity, this results in the cooling of the entire cathode and the solidification of melted regions (if any). After this, the heating—cooling process can theoretically be repeated. We did not observe such oscillations in numerical experiments, apparently, because of the restricted computational capability. But even without this, the effect of nonmonotone behavior of the free interface between the melt and the solid, which was observed in modeling of this system, seems very interesting even if we forget the "experimental" parameters given in [10]. From the standpoint of nonlinear equations, the free interface is the front of a nonlinear wave (solution of the Allen–Cahn equation, see (3.13)), and there are not so many known examples of nonmonotone behavior of nonlinear waves in real dissipative systems.

The numerical experiments permit separating the parameters that significantly influence the character of the process. Such parameters are the applied voltage, the rounding of the top of the cathode, and the parameters of the cathode base cooling.

The less significant influence is exerted by the combination of the parameters entering the dimensionless coefficient of thermal conductivity and the "form-factor" introduced to take into account the deviation due to the real geometry of the top of the cathode from the solutions of quantum mechanical problem of electron tunneling (emission of electrons from the cathode material) in the half-space.

1.4 Mathematical Statement of the Problem. Heat Transfer Model

Recall that the field emission of electrons is the phenomenon of electron emission in a vacuum under the action of an electric field.

To create a field of strength sufficient for the field emission in the case of a cathode shaped as a needle with the rounding radius of its top of the order of several nanometers, it is required to apply the voltage of only dozens or even several volts. But in the case of a cathode of small dimensions, there is a danger of its fracture (melting) under the action of the Joule heat generated by the current flowing through the cathode. In this section, we consider methods for modeling the heat transfer in such a small-size cathode. The dimensions of the silicon cathode used in physical experiments [10] are given in Table 1.1, and its image is shown in Fig.1.2.

The mathematical model of the heat transfer in the field emission process is well known, see, e.g., [20, Sect. 3.3.3]:

$$\rho c(T)\frac{\partial T}{\partial \tilde{t}} = \nabla\big(\lambda(T)\nabla T\big) + F, \tag{1.2}$$

$$\mathrm{div}\ \mathbf{j}_{\mathrm{in}} = 0. \tag{1.3}$$

Here $\tilde{t}$ is the real time, T is the temperature of the cathode material, ρ, c, and λ are the density, specific heat capacity, and the coefficient of heat conductivity of the cathode material, $\mathbf{j}_{\mathrm{in}}$ is the current density in the interior of the cathode, and F is the density of heat release power due to the Joule and Thomson effects. The function F has the form

$$F = |\mathbf{j}_{\mathrm{in}}|^{2}/\sigma_{\mathrm{e}}(T) + G(T)\langle\mathbf{j}_{\mathrm{in}}, \nabla T\rangle, \tag{1.4}$$

where $\sigma_{\mathrm{e}}(T)$ is the specific conductivity of the cathode and $G(T)$ is the Thomson coefficient. The current density is given by the formula

$$\mathbf{j}_{\mathrm{in}} = -\sigma_{\mathrm{e}}(T)\big(\nabla\Psi + A(T)\nabla T\big), \tag{1.5}$$

where Ψ is the electric field potential in the cathode and $A(T)$ is the coefficient of thermal electromotive force (EMF), see Sect. 2.3.

Table 1.1 Geometric parameters of the cathode [10]

Cathode height	10–15 μm
Cathode base diameter	6 μm
Rounding radius of the top of the cathode	15 nm
Angle at the top of the cathode	20°

We will consider a modification of this model adapted to the study of silicon cathodes of small dimensions. In this case, the Thomson effect can be neglected, because the silicon emitter has conductivity of electron-hole type and the contributions of electrons and holes to the thermal EMF approximately compensate each other, i.e., the coefficients $A(T) \approx 0$ in (1.5) and $G(T) \approx 0$ in (1.4), see Sect. 2.3 for details.

But this model is insufficient for the study of heat transfer accompanied by a phase transition (melting or crystallization). To describe the dynamics of interface in phase transitions, it is necessary to supplement heat equation (1.4) with the Stefan condition on the free boundary (interface) $\Gamma(\tilde{t})$:

$$k\left[\frac{\partial T}{\partial \mathbf{n}}\right]\Bigg|_{\Gamma(\tilde{t})} = v_{\mathbf{n}}, \tag{1.6}$$

where $\mathbf{n}$ is the outer normal on the free boundary and $k = \lambda/(c\rho)$. The normal is directed from the solid phase ($T = T_{\text{sol}}$) into the liquid phase ($T = T_{\text{liq}}$). In condition (1.6), the quantity $v_{\mathbf{n}}$ is the normal velocity of the free boundary $\Gamma(\tilde{t})$ and

$$\left[\frac{\partial T}{\partial \mathbf{n}}\right]\Bigg|_{\Gamma(\tilde{t})} = \langle \nabla T_{\text{sol}} - \nabla T_{\text{liq}}, \mathbf{n}\rangle.$$

Moreover, on the free boundary $\Gamma(\tilde{t})$, the Gibbs–Thomson condition must be satisfied:

$$(T - T_0)\big|_{\Gamma(\tilde{t})} = -\tilde{\alpha} v_{\mathbf{n}} - \tilde{\beta}\mathcal{K}, \tag{1.7}$$

where $\mathcal{K}$ is the average curvature of the free boundary, T_0 is the melting temperature of the cathode material, $\tilde{\alpha}$ and $\tilde{\beta}$ are constants determined by the physical parameters of the medium, namely, $\tilde{\alpha} = 1/\mu$, where μ is the kinetic growth coefficient, and $\tilde{\beta} = \sigma T_0/\rho l$, where σ is the surface tension, ρ is the density, and l is the latent heat of melting, see Table 4.1 in Sect. 4.1 for details. The law (1.7) describes the simplest (linear) dependence of the free boundary velocity on the temperature and curvature. We stress that, in the general case, it is necessary to use precisely condition (1.7) rather than the frequently encountered condition

$$T\big|_{\Gamma(\tilde{t})} = T_0. \tag{1.8}$$

Only in the case $\tilde{\beta} \ll 1$ and $\tilde{\alpha} \ll 1$, condition (1.7) implies (1.8).

Further, it is necessary to add the boundary condition determining the thermal balance in the thermo-field emission from the top of the cathode. It is described by the formula [20, Sect. 3.3.3]:

$$\lambda\frac{\partial T}{\partial \mathbf{n}}\bigg|_{S_e} = \frac{j_{\text{em}}}{e}\mathcal{E}_{\text{Nott}}\bigg|_{S_e} - \psi\sigma_{\text{SB}}T^4\bigg|_{S_e}. \tag{1.9}$$

Here $\mathbf{n}$ is the normal on the emitting electron surface S_e directed into a vacuum, σ_{SB} is the Stefan–Boltzmann constant, e is the electron charge, j_{em} is the emission current density value, $\mathcal{E}_{\mathrm{Nott}}$ is the average energy of the Nottingham effect (see Sect. 2.4.7), and ψ is the degree of blackness (the radiation coefficient that is the ratio of the thermal radiation energy of a "gray body" to that of an "absolutely black body" at the same temperature, $0 \leqslant \psi < 1$). The first term in the right-hand side of (1.9) corresponds to the Nottingham effect (see Sect. 2.4.7), and the second term corresponds to the radiation by the Stefan–Boltzmann law, which because of a small value of the Stefan–Boltzmann constant σ_{SB} (see Table 4.1 in Sect. 4.1) is small as compared to the term corresponding to the Nottingham effect. Therefore, we omit this term in calculations and use a boundary condition of the form

$$\lambda \frac{\partial T}{\partial \mathbf{n}}\bigg|_{S_{\mathrm{e}}} = \frac{j_{\mathrm{em}}}{e}\mathcal{E}_{\mathrm{Nott}}\bigg|_{S_{\mathrm{e}}} \stackrel{\text{def}}{=} \lambda f_{\mathrm{Nott}}(T, j_{\mathrm{em}}, E_{\mathrm{F}}). \tag{1.10}$$

Thus, Eqs. (1.2)–(1.7) and (1.9) determine the model used in our study of heat transfer and field emission in nanocathodes. It is clear that they must be supplemented with the initial and boundary conditions on the outer boundaries of the cathode, and this will be done in Chap. 4.

References

1. Becquerel, A.E.: Mémoire sur les effets électriques produits sous l'influence des rayons solaires. Comptes Rendus **9**, 561–567 (1839)
2. Bonch-Bruevich, V.L., Kalashnikov, S.G.: Semiconductor Physics. Nauka, Moscow (1990). (in Russian)
3. Bugaev, S.P., Litvinov, E.A., Mesyats, G.A., Proskurovskii, D.I.: Explosive emission of electrons. Sov. Phys. Uspekhi **18**, 51–61 (1975)
4. Cathey, D.A.: Field-emission displays. Inf. Disp. **16** (1995)
5. Christov, S.G.: General theory of electron emission from metals. Phys. Status Ssolidi (B) **17**(1), 11–26 (1966)
6. Danilov, V.G.: On the relation between the Maslov-Whitham method and the weak asymptotics method. In: Kamiński, A., Oberguggenberger, M., Pilipović, S. (eds.) Linear and Non-Linear Theory of Generalized Functions and its Applications, vol. 88, pp. 55–65. Banach Center Publications, Warsaw (2010)
7. Danilov, V.G., Omel'yanov, G.A., Radkevich, E.V.: Asymptotic behavior of the solution of a phase field system, and a modified stefan problem. Differ. Equ. **31**(3), 446–454 (1995)
8. Danilov, V.G., Omel'yanov, G.A., Radkevich, E.V.: Hugoniot-type conditions and weak solutions to the phase-field system. Eur. J. Appl. Math. **10**, 55–77 (1999)
9. Danilov, V.G., Omel'yanov, G.A., Shelkovich, V.M.: Weak asymptotics method and interaction of nonlinear waves. Am. Math. Soc. Transl.: 2 **208**, 33–163. Providence: American Mathematical Society (2003)
10. Dyzhev, N.A., Gudkova, S.A., Makhiboroda, M.A., Fedirko V, A.: Investigation of emussion properties of silicon cathodes of different geometry. In: Bykov, D.V. (Ed.) Vacuum science and Technics, Material of XII Scientific-Technical Conference with Participation of Foreign Specialists, pp. 221–224. MIEM, Moscow (2005). (in Russian)
11. Egorov, N., Sheshin, E.: Field Emission Electronics. Springer (2017)
12. Elster, G.: On the electricity of flames. Annalen of Physik und Chemie **3**(16), 193–222 (1882)
13. Elster, G.: On the generation of electricity by the contact of gases and incandescent bodies. Annalen of Physik und Chemie **3**(19), 588–624 (1883)

14. Elster, G.: On the unipolar conductivity of heated gases. Annalen of Physik und Chemie **3**(26), 1–9 (1885)
15. Elster, G.: On the electrification of gases by incandescent bodies. Annalen of Physik und Chemie **3**(31), 109–127 (1887)
16. Elster, G.: On the generation of electricity by contact of rarefied gas with electrically heated wires. Annalen of Physik und Chemie **3**(37), 315–329 (1889)
17. Flügge, S. (ed.): Electron-Emission and Gas Discharges I, Encyclopedia of Physics, vol. XXI. Springer, Berlin (1956)
18. Fowler, R.H., Nordheim, L.: Electron emission in intense electric fields. Proc. Royal Soc. Lond. Ser. A **119**(781), 173–181 (1928)
19. Furcey, G.: Field Emission in Vacuum Microelectronics. Springer (2005)
20. Goldstein, E.: On electric conduction in vacuum. Annalen der Physik und Chemie **3**(24), 79–92 (1885)
21. Grundmann, M.: The Physics of Semiconductors: An Introduction Including Nanophysics and Applications. Springer (2016)
22. Guthrie, F.: On a relation between heat and static electricity. Philos. Mag. **46**(306), 257–266 (1873)
23. Hertz, H.: Ueber einen einfluss des ultravioletten lichtes auf die electrische entladung. Annalen der Physik **267**(8), 983–1000 (1887)
24. Hittorf, W.: On electrical conduction of gases. Annalen of Physik und Chemie **2:136**, 1–31, 197–234 (1869)
25. Hittorf, W.: On electrical conduction of gases. Annalen of Physik und Chemie Jubalband **430–445** (1874)
26. Hittorf, W.: On electrical conduction of gases. Annalen of Physik und Chemie **3**(7), 553–631 (1879)
27. Hittorf, W.: On electrical conduction of gases. Annalen of Physik und Chemie **3**(20), 705–775 (1883)
28. Hittorf, W.: On electrical conduction of gases. Annalen of Physik und Chemie **3**(21), 90–139 (1884)
29. Hofmann, P.: Solid State Physics: An Introduction. Wiley (2015)
30. Lee, T.H.: T-f theory of electron emission in high-current arcs. J. Appl. Phys. **30**(2), 166–171 (1959)
31. Mironov, V.L.: Fundamentals of Scanning Probe Microscopy. Institute for Physics of Microstructures RAS, Nizhniy Novgorod (2004)
32. Murphy, E.L., Good, R.H.: Thermionic emission, field emission, and the transition region. Phys. Rev. **102**(6), 1464–1473 (1956)
33. Nottingham, W.B.: Remarks on energy losses attending thermionic emission of electrons from metals. Phys. Rev. (1941)
34. Paulini, J., Klein, T., Simon, G.: Thermo-field emission and the Nottingham effect. J. Phys. D: Appl. Phys. **26**(8), 1310–1315 (1993)
35. Richardson, O.: Thermionic phenomena and the laws which govern them. In: Nobel Lecture, pp. 224–236. Stockholm (1929)
36. Schottky, W.: Über kalte und warme elektronenentladungen. Zeitschrift für Physik **14**(1), 63–106 (1923)
37. Shalimova, K.V.: Physics of Semiconductors. Energoatomizdat, Moscow (1985). (in Russian)
38. Smith, R.T.: Electronics developments for field-emission displays. Inf. Display **14**(2), 12 (1998)
39. Smith, W.: Effect of light on selenium during the passage of an electric current. Nature **7**(173), 303 (1873)
40. Stilbans, L.S.: Physics Semiconductors. Soviet Radio, Moscow (1967). (in Russian)
41. Stoletow, A.: Suite des recherches actino-electriques. Comptes Rendus CVII **91** (1888)
42. Stoletow, A.: Sur les courants actino-electriqies au travers detair. Comptes Rendus CVI **1593** (1888)
43. Stoletow, A.: Sur une sorte de courants electriques provoques par les rayons ultraviolets. Comptes Rendus CVI **1149** (1888)
44. Stratton, R.: Theory of field emission from semiconductors. Phys. Rev. **125**(1), 67–82 (1962)

Chapter 2
Physical Basis for Field Emission

Abstract In this chapter, without claiming to be original, we recall some basic notions of the solid-state physics which will be used to construct the mathematical model of the field emission cathode. More detailed descriptions of these facts can be found in any literature on the solid-state physics. Our description is based on [2, 3, 23, 27, 28, 30, 41, 42]. In this chapter, we also present some known notions from the theory of field emission based on the papers and monographs [4, 10–12, 17, 20, 34, 36, 37, 39, 43] and several others which will be mentioned in the course of presentation.

2.1 Energy-Band Theory and Fermi Level

It is well known that the field emission can be observed from both metallic and semiconducting emitters. In our book, we model the field emission from silicon.

We first consider the energy-band structure, because the energy-band theory underlies the contemporary theory of solids, and precisely this theory allows one to understand the nature of conductors (metals) and semiconductors and to explain their important properties. By the quantum mechanics, the free electrons can have any energy [32], because their energy spectrum is continuous. But by the Bohr postulates, the electrons in isolated atoms can have only some definite discrete values of the energy. In this case, the electron is said to be in one of the orbitals. In a solid, the energy spectrum is different; it consists of separate permitted energy bands separated by band gaps. In the case of several atoms united by chemical bonds, for example, in a molecule, the number of electron orbitals is proportional to the number of atoms. In crystals, the number of orbitals is very large, and the difference between the energies of electrons in neighboring orbitals is very small, i.e., the energy levels split and practically fill several intervals (energy bands). There are three bands: conduction band, valence band, and band gap. The valence band is the lowest band, where at the temperature $0\,\text{K}$, all energy states are occupied by electrons.

© Springer Nature Singapore Pte Ltd. 2020

V. Danilov et al., *Mathematical Modeling of Emission in Small-Size Cathode*,
Heat and Mass Transfer, https://doi.org/10.1007/978-981-15-0195-1_2

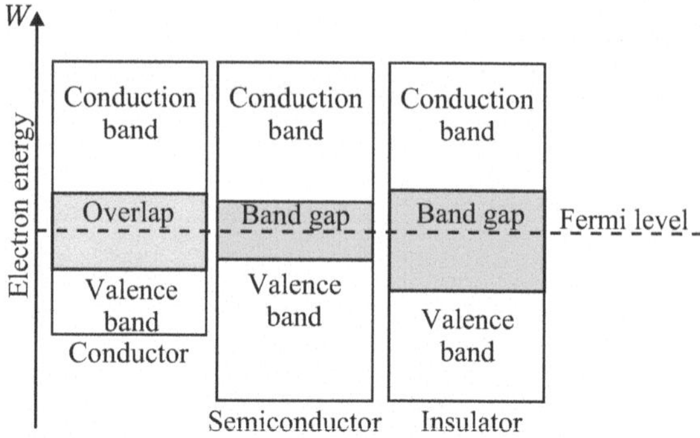

Fig. 2.1 Energy-band structure of solids

The mutual arrangement of these bands is one of the main differences between metals and semiconductors. In metals (conductors), the conduction band and the valence band overlap thus forming one band while the band gap is absent. Thus, an electron can freely move between the bands and acquire any admissible small energy. And in semiconductors, the valence band and the conduction band do not overlap and the band gap is located between them (see Figs. 2.1 and 2.2a). To take an electron from the valence band into the conduction band in the pure semiconductor (i.e., in the semiconductor that completely consists of atoms of one element, for example, silicon, without inclusions of atoms of other elements), it is required to apply more energy than in metals.

But this can be changed by adding an admixture of a different substance to pure semiconductors. This process is called the doping, and the obtained semiconductor is said to be extrinsic. In the obtained doped semiconductor, the admixture creates additional energy levels in the band gap on the side of the conduction band (donor impurity) or, conversely, on the side of the valence band (acceptor impurity) thus decreasing the minimal energy necessary for the electron transition from the valence band to the conduction band (Fig. 2.2b, c), see for details below.

In Fig. 2.2a, by W_c we denote the energy of the lower level in the conduction band or, as one usually says, the bottom energy of the conduction band, by W_v we denote the energy of the highest level of the valence band or, as one usually says, the ceiling energy of the valence band, and by W_g we denote the width of the band gap which is equal to $W_g = W_c - W_v$.

The width of the band gap in semiconductors is not constant but depends on the temperature. At lower temperatures, it expands and the semiconductor thus practically becomes a dielectric, and as the temperature increases, it becomes narrower and the properties of semiconductor thus approach the properties of metals.

The width of the band gap W_g is approximately described by the Varshni relation [44, 46]

$$W_g(T) = W_0 - \frac{\alpha_1 T^2}{\alpha_2 + T},$$

(2.1)

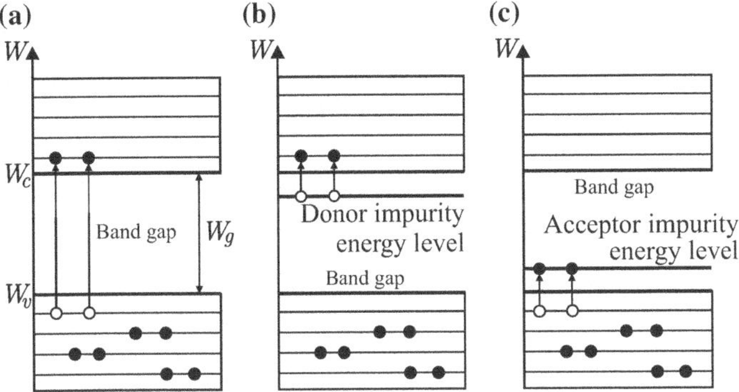

Fig. 2.2 Energy-band structure of semiconductors ($T > 0$ K): **a** pure semiconductor, **b** and **c** extrinsic semiconductors

where W_0 is the width of the band gap at the temperature $T = 0$ K and α_1, α_2 are constants that, in the case of silicon, are equal to $W_0 = 1.17$ eV, $\alpha_1 = 7.021 \times 10^{-4}$ eV/K, and $\alpha_2 = 1108$ K. We note that the second term in the right-hand side of (2.1) is much less than W_0. Therefore, the width of the band gap $W_g(T)$ changes little as the temperature T increases.

Now let us consider the mechanism of electric conductivity of semiconductors from the standpoint of the crystal structure. In a crystal, the atoms of the crystal lattice of semiconductors form covalent bonds. The covalent bond between two atoms is a chemical bond formed by a pair of overlapping valence electron clouds. The electrons bonding two atoms are called a shared electron pair.

For example, each atom in silicon is surrounded by four neighbors and is connected with each of them by the covalent bond (see Fig. 2.3). As a result, a cubic crystal lattice is formed which is called a diamond-type lattice and is characterized by the lattice constant a_{Si}.

Generally speaking, in a semiconductor of ideal structure, all electrons are in a bound state (see Fig. 2.4a). We assume that, under the action of some perturbations, for example, the temperature, a covalent bond breakage occurred in the semiconductor (see Fig. 2.4b) and a bound electron became free. This process is called generation. After the electron leaves the covalent bond, it has an excessive positive charge, i.e., there is a vacant place which is called a hole. A free electron can occupy this vacant place and become a bound electron. This process is called recombination. The hole arising in generation can be filled with a free electron from a neighboring covalent bond. Then a hole appears in the bond from which the electron arrived, i.e., the hole can travel.

In the absence of an electric field, the velocities of free electrons can arbitrarily be directed, because they make only the thermal motion. Since the thermal motion is random, the mean value of the thermal velocity is zero (there are no preferred directions). This means that each moving electron must be associated with a hole moving in the opposite direction with the same velocity, or conversely, each moving

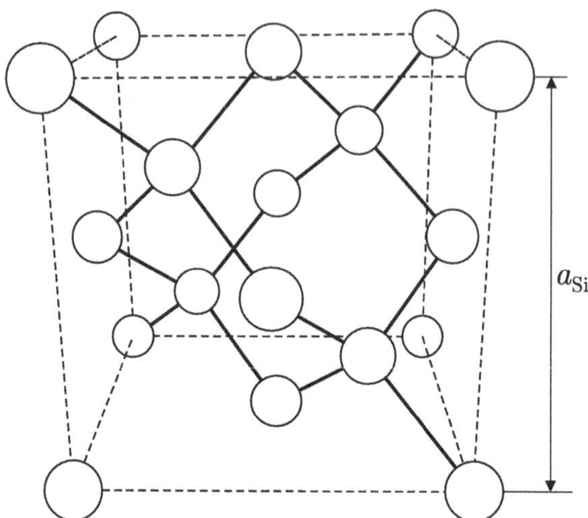

Fig. 2.3 Crystal lattice of silicon

hole must be associated with an electron. Therefore, the number of free electrons and holes moving in a certain direction is approximately equal to the number of electrons and holes moving in the opposite direction. If the semiconductor is located in an external electric field, then the electrons move against the field direction, and the holes move in the field direction, which is equivalent to the motion of a positive charge along the field. This mechanism is called the electron–hole conduction. From the standpoint of the energy-band theory, this is equivalent to the motion of an electron from the valence band to the conduction band (generation) and conversely (recombination).

The doping means that several atoms at crystal lattice sites are replaced by atoms of different chemical elements. The following two types of such replacements are possible. The first type is the replacement by an atom of greater valence. This means that, after formation of bounds with the neighboring elements of the crystal lattice, the impurity atom has one unoccupied valence electron which, under the action of a perturbation (for example, temperature) becomes free. This is a donor impurity, and the obtained semiconductor is an n-type semiconductor. The second type is the replacement by an atom with a lesser valence. In this case, after formation of bonds with the neighboring elements, a part of bonds remains incomplete (because the number of valence electrons of the impurity atom is insufficient for the formation of all bonds), i.e., there arise holes that can be occupied by free electrons. This is an acceptor impurity, and the obtained semiconductor is a p-type semiconductor.

The behavior of the system of electrons in a solid is described by the Fermi–Dirac statistics. The probability of the presence of electrons at the level of energy W is described by the function

(a)   **(b)**

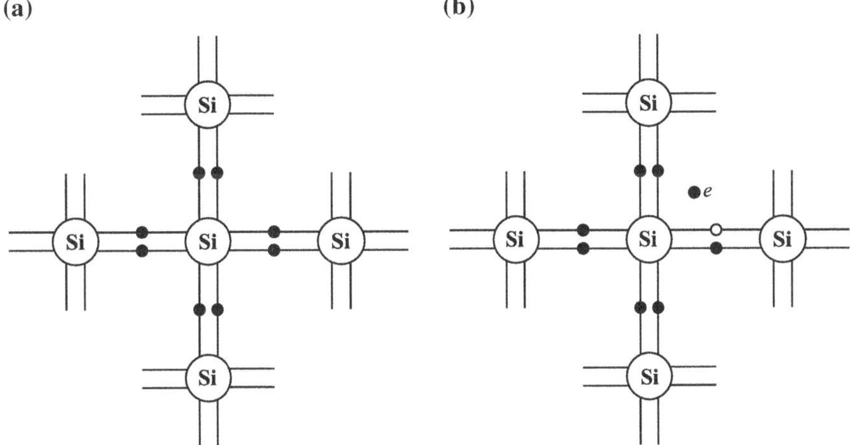

Fig. 2.4 Electric conductivity mechanism in silicon. The black circles are electrons, the white circle is a hole

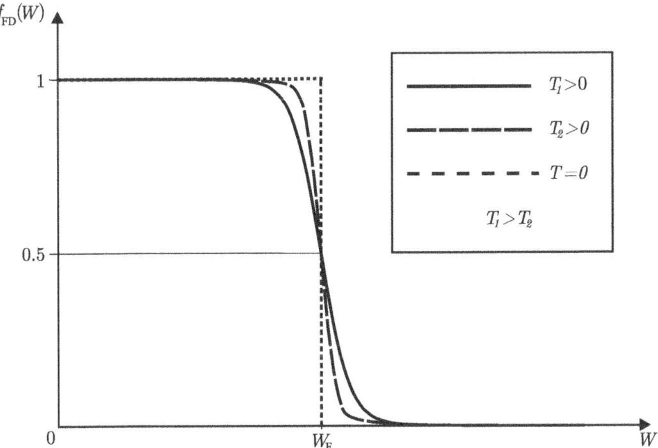

Fig. 2.5 Graph of the Fermi–Dirac function

$$f_{FD}(W) = \left[\exp\left(\frac{W - W_F}{k_B T} \right) + 1 \right]^{-1}, \tag{2.2}$$

where k_B is the Boltzmann constant, T is the absolute temperature, and W_F is the Fermi level. The graph of this function for different temperatures is shown in Fig. 2.5. The Fermi level has the following physical meaning: the probability of finding a particle at the Fermi level is $1/2$ for any temperatures except for $T = 0$.

But, in general, instead of the Fermi level in formula (2.2), there must be another physical quantity, i.e., the chemical potential μ_F, which is a thermodynamic function determining the variation in thermodynamic potentials (internal energy, etc.) as the number of particles in the system varies. But at the temperatures less than the characteristic Fermi temperature W_F/k_B, which is of the order of 10^5 K (the Boltzmann constant is $k_B = 8.617 \times 10^{-5}$ eV·K^{-1}, and the Fermi energy in silicon W_F is or the order of several electronvolts and changes little as the temperature varies), one can assume that the chemical potential is $\mu_F \approx W_F$.

It should be noted that, for the energies $W - W_F > 2k_B T$, the Fermi–Dirac distribution function can be replaced by the Maxwell–Boltzmann distribution function $f_{MB}(W)$:

$$f_{FD}(W) \approx \exp\left(-\frac{W - W_F}{k_B T}\right) \stackrel{\text{def}}{=} f_{MB}(W). \tag{2.3}$$

In the pure semiconductors, the Fermi level is described by the expression

$$W_F = W_i + \frac{3}{4}k_B T \ln \frac{m_n^*}{m_p^*}, \tag{2.4}$$

where m_n^* and m_p^* are the effective masses of electrons and holes (see about them below) and W_i is the energy in the middle of the band gap which has the form (see Fig. 2.2a)

$$W_i = W_v + \frac{W_g}{2} = \frac{W_v + W_c}{2}.$$

The second term in formula (2.4) is negligibly small, and hence one can say that the Fermi level in pure semiconductors lies in the middle of the band gap:

$$W_F \approx W_i.$$

The doping can shift the energy level up and down, but it still remains in the band gap.

2.2 Conductivity of Semiconductors

Now we consider one of the most important properties of solids, the conductivity σ_e (or the specific resistance ρ_e which is inversely proportional to the conductivity $\rho_e = 1/\sigma_e$). Here and below, we consider an ideal crystal, i.e., a semiconductor whose crystal lattice does not have any defects (for example, the absence of an atom at any site of the crystal lattice, its deformation in a certain direction, and so on).

As we already noted, in the absence of an electric field, the velocities of free electrons have all possible directions because of their thermal motion. Since the thermal motion is random, the mean value of the thermal velocity is zero (there are no preferred directions) which means that the mean current density is also zero.

In the presence of an external field, the electrons acquire additional velocities under the action of the field. In this case, the motions of electrons cease to be random, i.e., the resultant direction is distinguished and an electric current (directed flow of electric charge) arises. The mean velocity of ordered motion $\mathbf{v}_d$ is called the drift velocity.

The drift velocity is proportional to the electric field strength:

$$\mathbf{v}_d = \mu_e \mathbf{E},$$

where μ_e is the proportionality coefficient called the carrier mobility which, by definition, is the drift velocity acquired by a particle in the field of strength 1. For electrons, the mobility is negative, and for positive particles, it is positive.

The density of electric current is

$$\mathbf{j} = en\mathbf{v}_d = en_e\mu_e\mathbf{E},$$

where e is the charge of a particle and n_e is the particle concentration. On the other hand, by Ohm's law,

$$\mathbf{j} = \sigma_e\mathbf{E},$$

where σ_e is the conductivity. This implies the formula

$$\sigma_e = en_e\mu_e. \tag{2.5}$$

In the case of semiconductors, the conduction obeys the electron-hole mechanism, and hence formula (2.5) becomes

$$\sigma_e = e(n\mu_n + p\mu_p), \tag{2.6}$$

where n and μ_n are the concentration and mobility of electrons and p and μ_p are the concentration and mobility of holes.

In the case of a pure semiconductor, $n = p$, and the intrinsic concentration of carriers n_i is

$$n_i^2 = np. \tag{2.7}$$

In this case, formula (2.6) becomes simpler:

$$\sigma_e = en_i(\mu_n + \mu_p). \tag{2.8}$$

2.2.1 Electron and Hole Concentration

The electron concentration in the conduction band n can be calculated by the formula

$$n = \int_{W_c}^{W_c^{\max}} f_{\mathrm{FD}}(W)\, \mathrm{d}n_{\mathrm{dens}}(W), \tag{2.9}$$

where $n_{\mathrm{dens}}(W)$ is the density of allowed states for electrons per unit energy interval, W_c is the bottom energy of the conduction band, see Fig. 2.2a, and $W_c^{\max}$ is the ceiling energy of the conduction band. Since the function $f_{\mathrm{FD}}(W)$ decreases fast (tends to zero) as W increases (see Fig. 2.5), we see that, in formula (2.9), the upper limit can be set equal to infinity. The hole concentration p is calculated by the formula

$$p = \int_{W_v^{\min}}^{W_v} \left(1 - f_{\mathrm{FD}}(W)\right) \mathrm{d}p_{\mathrm{dens}}(W), \tag{2.10}$$

where $p_{\mathrm{dens}}(W)$ is the maximal density of allowed states for holes, W_v is the ceiling energy of the valence band, and $W_v^{\min}$ is the bottom energy for the valence band (see Fig. 2.2a). Since as in the case of formula (2.9), the function $1 - f_{\mathrm{FD}}(W)$ tends to zero very fast as W decreases, the lower limit in formula (2.10) can be set equal to $-\infty$.

Now we derive expressions for $n_{\mathrm{dens}}(W)$ and $p_{\mathrm{dens}}(W)$). Generally speaking, to determine these functions in explicit form is a very difficult problem. But if we take into account that the function $f_{\mathrm{FD}}(W)$ decreases very sharp as W increases, then it suffices to know the expression for the function $n_{\mathrm{dens}}(W)$ near the bottom of the conduction band. Similarly, it suffices to know the expression for the function $p_{\mathrm{dens}}(W)$ near the ceiling of the valence band.

Let us determine the function $n_{\mathrm{dens}}(W)$ near the bottom of the conduction band. We distinguish the unit volume in the crystal. The electron energy near the bottom of the conduction band is described by the expression

$$W = W_c + \frac{\hat{p}^2}{2m_n^*} = W_c + \frac{\hat{p}_x^2}{2m_n^*} + \frac{\hat{p}_y^2}{2m_n^*} + \frac{\hat{p}_z^2}{2m_n^*}, \tag{2.11}$$

where m_n^* is the effective mass of electron and $\hat{p}$ is the quasimomentum.

We distinguish a spherical layer between two isoenergy surfaces (see Fig. 2.6):

$$W = \mathrm{const} \quad \text{and} \quad W + dW = \mathrm{const}.$$

The volume of this layer is equal to

$$\mathrm{d}V_{\hat{p}} = 4\pi \hat{p}^2 \mathrm{d}\hat{p}.$$

The volume of an elementary cell in the Brillouin zone, i.e., in the space of quasi-momenta of crystal of unit volume is equal to h^3 [23, 27, 41], where h is the Planck constant. By the *Pauli principle* [32], in each of such cells, there can be two elec-

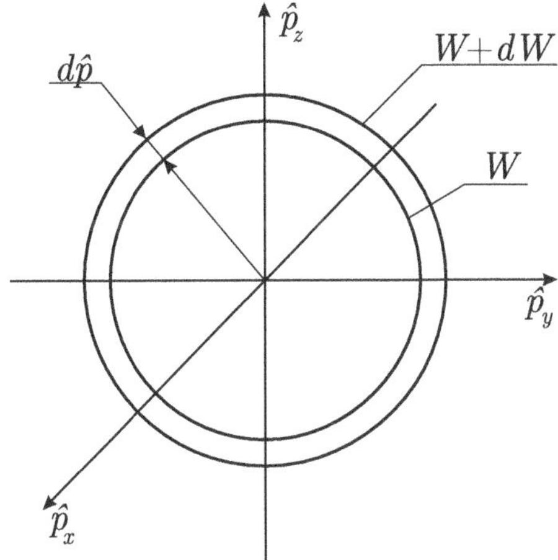

Fig. 2.6 Spherical layer between two isoenergy surfaces

trons (with oppositely directed spins). Therefore, the number of quantum states in the volume $dV_{\hat{p}}$ is equal to

$$\mathrm{d}n_{\mathrm{dens}} = 2\mathrm{d}V_{\hat{p}}/h^3 = 2 \cdot 4\pi\hat{p}^2\mathrm{d}\hat{p}/h^3. \tag{2.12}$$

From (2.11) we derive the formulas

$$\mathrm{d}\hat{p} = m_n^* dW/\hat{p}; \quad \hat{p}^2 = 2m_n^*(W - W_c). \tag{2.13}$$

Substituting (2.13) into (2.12), we obtain

$$\mathrm{d}n_{\mathrm{dens}} = 2\frac{4\pi 2m_n^*(W - W_c)m_n^* dW}{\sqrt{2m_n^*(W - W_c)}h^3} = \frac{4\pi(2m_n^*)^{\frac{3}{2}}}{h^3}\sqrt{(W - W_c)}\mathrm{d}W, \tag{2.14}$$

Similarly, we can obtain the formula for $p_{\mathrm{dens}}(W)$:

$$\mathrm{d}p_{\mathrm{dens}}(W) = \frac{4\pi(2m_p^*)^{\frac{3}{2}}}{h^3}\sqrt{W_v - W}\,\mathrm{d}W, \tag{2.15}$$

where m_p^* is the effective mass of holes (which is different from the effective mass of electrons m_n^*).

We note that if the isoenergy surfaces were ellipsoids rather than spheres, and this is possible if

$$W = W_c + \frac{\hat{p}_x^2}{2m_1^*} + \frac{\hat{p}_y^2}{2m_2^*} + \frac{\hat{p}_z^2}{2m_3^*},$$

then proceeding in the same way and taking the effective mass as

$$m_n^* = (m_1^* m_2^* m_3^*)^{1/3},$$

we would obtain precisely the same formula (2.14). A similar formula can also be written for the effective mass of holes m_p^*, which finally also leads to formula (2.15).

Substituting expressions (2.14) and (2.3) into (2.9), we obtain the formula for the electron concentration:

$$n = \frac{4\pi(2m_n^*)^{\frac{3}{2}}}{h^3} \int_{W_c}^{\infty} \sqrt{W - W_c} \exp\left(-\frac{W - W_F}{k_B T}\right) dW$$

$$= N_c \exp\left(\frac{W_F - W_c}{k_B T}\right), \tag{2.16}$$

where

$$N_c = 2\frac{(2\pi m_n^* k_B T)^{\frac{3}{2}}}{h^3}. \tag{2.17}$$

Similarly, we obtain the formula for the hole concentration:

$$p = \frac{4\pi(2m_p^*)^{\frac{3}{2}}}{h^3} \int_{-\infty}^{W_v} \sqrt{W_v - W} \exp\left(\frac{W - W_F}{k_B T}\right) dW$$

$$= N_v \exp\left(\frac{W_v - W_F}{k_B T}\right), \tag{2.18}$$

where

$$N_v = 2\frac{(2\pi m_p^* k_B T)^{\frac{3}{2}}}{h^3}. \tag{2.19}$$

The coefficients N_c and N_v are called *effective densities of states*.

In pure semiconductors, as was already noted, $n = p = n_i$, i.e., $np = n_i^2$ (see (2.7)). With $W_g = W_c - W_v$ in mind, we obtain

$$n_i = \sqrt{np} = \sqrt{N_c N_v} \exp\left(\frac{-W_g}{2k_B T}\right). \tag{2.20}$$

2.2.2 Effective Mass

The notion of effective mass was used above. By definition, the *effective mass of a particle* is the dynamic mass which arises when the particle moves in the periodic potential of the crystal. This notion is introduced to simplify the mathematical description of motion of carriers in the potential field of crystal lattice. The charge carriers in the crystal interact with the electric field as if they were freely moving in a vacuum but with a certain effective mass. The effective mass is determined by the expression

$$m^* = \hbar^2 \left(\frac{d^2 W(k)}{dk^2} \right)^{-1},$$ (2.21)

where $W(k)$ is the dispersion law. In the crystal, the effective mass is different from the mass at rest, but for a free particle, the effective mass is equal to the mass at rest. We also note that the effective mass of a hole differs from the effective mass of an electron (for example, in silicon at $T = 4.2\,\text{K}$, the effective mass of an electron is equal to $m_n^* = 1.06\,m_e$, and the effective mass of a hole is equal to $m_p^* = 0.59\,m_e$, where m_e is the rest mass of an electron and $m_e = 9.11 \times 10^{-31}\,\text{kg}$). It is not constant for a crystal and varies depending on the temperature.

2.2.3 Electron and Hole Mobilities

Now we consider how the mobility depends on the temperature. By definition, $\mu_e = \mathbf{v}_d / \mathbf{E}$. The drift velocity can be expressed in terms of the free path time (relaxation time) as $\mathbf{v}_d = e\mathbf{E}\tau / m^*$, where τ is the free path time and m^* is the effective mass. We obtain the formula for μ_e:

$$\mu_e = \frac{e\tau}{m^*}.$$ (2.22)

The relaxation time τ is equal to the ratio of the free path length to the speed of thermal motion of the carrier: $\tau = \lambda_e / v_T$, where λ_e is the free path length and v_T is the thermal velocity which has the form

$$v_T = \sqrt{\frac{3k_B T}{m^*}}.$$

In the general case, the problems of calculating the relaxation time and the free path length are very difficult. Under our assumption (we consider an ideal crystal), we need to determine the free path length λ_e only in the case of carrier scattering at the lattice vibrations, which has the form [42] $\lambda_e \approx A_e / T$, where the coefficient A_e depends on the matter and the type of carriers. But if the crystal is imperfect, then it is also necessary to consider the scattering at impurities, the scattering at defects, and so on.

As a result, the following formulas for the mobilities are usually used:

$$\mu_n \approx \frac{e A_n}{\sqrt{3k_B m_n^*}} T^{-3/2}; \qquad \mu_p \approx \frac{e A_p}{\sqrt{3k_B m_p^*}} T^{-3/2}.$$ (2.23)

Fig. 2.7 Approximate graph of $\sigma_e(T)$ for silicone (in SI units)

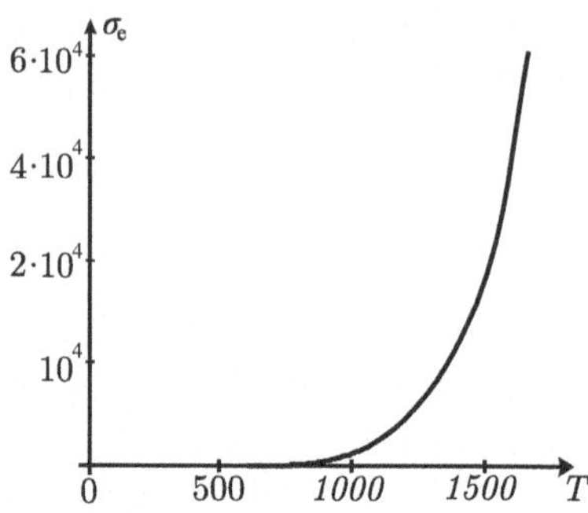

2.2.4   Temperature-Dependence of Conductivity in Silicon

Finally, with regard to (2.20) and (2.23), we see that formula (2.8) becomes

$$\sigma_e(T) = \sigma_0 \exp\left(\frac{-W_g}{2k_B T}\right), \tag{2.24}$$

where

$$\sigma_0 = e\sqrt{N_c N_v}\left(\frac{eA_n}{\sqrt{3k_B m_n^*}}T^{-3/2} + \frac{eA_p}{\sqrt{3k_B m_p^*}}T^{-3/2}\right)$$

$$= e\frac{(2\pi k_B)^{\frac{3}{2}}}{h^3}\sqrt{(m_n^* m_p^*)^{\frac{3}{2}}}\left(\frac{eA_n}{\sqrt{3k_B m_n^*}} + \frac{eA_p}{\sqrt{3k_B m_p^*}}\right).$$

Figure 2.7 shows an approximate graph of the function $\sigma_e(T)$ for silicon under the assumption that the effective masses of electrons and holes (m_n^* and m_p^*) are independent of the temperature. In reality, they nonlinearly depend on the temperature, but the value of their change to the melting temperature is small, and to determine this dependence is a rather complicated problem.

Note that formula (2.24) is valid for $T < T_0$, where T_0 is the melting point temperature, and after melting, the conductivity has a jump, see Sect. 4.8, Fig. 4.31 and [21] for more details.

2.3   Thermoelectricity

The *thermoelectric phenomena* are all the phenomena due to the relationship between thermal and electrical processes in metals and semiconductors. In the case of field emission which we consider, the Thomson effect is possible.

One of the thermoelectric phenomena, the *Thomson effect*, is that, in a homogeneous conductor which is inhomogeneously heated by a direct current, in addition to the heat released according to the Joule–Lentz law

$$F_{\mathrm{JL}} = |\mathbf{j}_{\mathrm{in}}|^2 / \sigma_{\mathrm{e}}(T), \tag{2.25}$$

some additional Thomson heat F_{T} is released or absorbed in the conductor volume depending on the current direction (i.e., if the temperature gradient coincides with the current direction, then the heat is released, and if they are oppositely directed, then the heat is absorbed):

$$F_{\mathrm{T}} = G(T)\langle \mathbf{j}_{\mathrm{in}}, \nabla T \rangle, \tag{2.26}$$

where $G(T)$ is the Thomson coefficient, $\mathbf{j}_{\mathrm{in}}$ is the current density inside the cathode, and T is the cathode temperature.

The *thermal electromotive force (EMF)*, i.e., the appearance of a potential difference at the ends of a homogeneous conductor in the case of a temperature difference at its ends, is also possible. The density of the thermal EMF current $\mathbf{j}_{\mathrm{te}}$ is given by the formula

$$\mathbf{j}_{\mathrm{te}} = -\sigma_{\mathrm{e}}(T)A(T)\nabla T, \tag{2.27}$$

where $A(T)$ is the thermal EMF coefficient.

The temperature-dependence of the thermal EMF of a semiconductor, $A(T)$, is approximately described by Pisarenko's formula

$$A(T) = -\frac{k_{\mathrm{B}}}{\sigma_{\mathrm{e}}(T)}\left(\mu_n n\left[C + \ln\frac{N_c}{n}\right] - \mu_p p\left[C + \ln\frac{N_v}{p}\right]\right), \tag{2.28}$$

where C is a constant whose value depends on the electron scattering mechanism; for a pure semiconductor, $C = 2$ [42].

In the pure semiconductor, this formula is simpler:

$$A(T) = -\frac{k_{\mathrm{B}}}{e}\frac{\mu_n - \mu_p}{\mu_n + \mu_p}\left[C + \frac{W_g}{2k_{\mathrm{B}}T}\right] - \frac{k_{\mathrm{B}}}{e}\ln\sqrt{\frac{(m_n^*)^{3/2}}{(m_p^*)^{3/2}}}. \tag{2.29}$$

The graph of the function $A(T)$ in (2.29) is shown in Fig. 2.8 (for silicon). In this case, it was also assumed that m_n^* and m_p^* are independent of the temperature (see above for details).

The Thomson coefficient $G(T)$ can be expressed in terms of thermal EMF by the formula

$$G(T) = -T\frac{\mathrm{d}A(T)}{\mathrm{d}T}. \tag{2.30}$$

The graph of the function $G(T)$ for silicon is shown in Fig.2.9 (under the assumption that m_n^* and m_p^* are independent of the temperature, see above for details).

The graphs presented in this section (see Figs. 2.8 and 2.9) show that, in semiconductors, the thermoelectric phenomena are manifested extremely weakly and can be

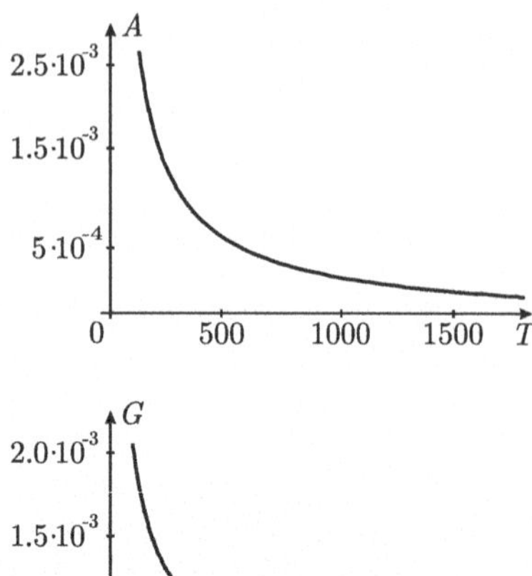

Fig. 2.8 Thermal EMF coefficient versus the temperature in silicon

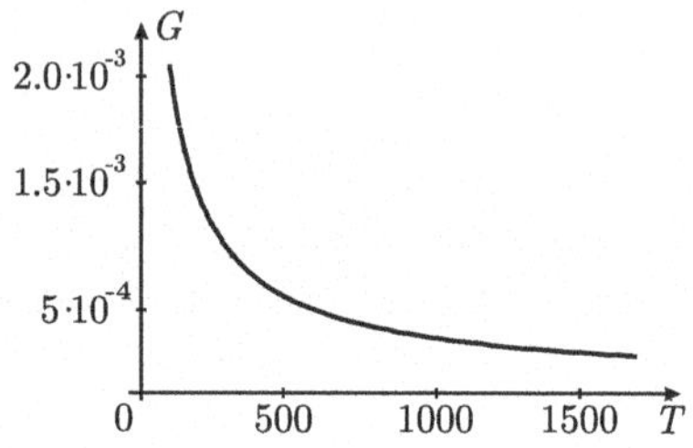

Fig. 2.9 Thomson coefficient versus the temperature in silicon

neglected. This is explained by the fact that the current carriers in semiconductors, i.e., electrons and holes, approximately compensate each other.

2.4 Emission Current Density and Nottingham Effect

As was previously mentioned (see Sect. 1.2), the thermal electron emission from a solid arises when it is heated to high temperatures and is located in a field of small strength. This phenomenon has been studied well. The density of emission current in the thermionic emission j_T (in emission from a metal to a vacuum) is described by the Richardson formula [40]

$$j_T = (1 - R)A_0 T^2 \exp\left(-\frac{\varsigma}{k_B T}\right), \tag{2.31}$$

where T is the emitter temperature, k_B is the Boltzmann constant, A_0 is the thermionic constant,

$$A_0 = \frac{4\pi m_e k_B^2 e}{h^3},$$

m_e and e are the electron mass and charge, R is the mean value of the electron reflection from the potential barrier, and ς is the work function, i.e., the difference between the maximal energy required by an electron to be deleted from the solid (and to be transferred to a point that is sufficiently far from the surface on the atomic scale) and the Fermi level energy. The modification of this formula that includes the Schottky effect, i.e., the phenomenon of a decrease in the barrier when the emitter is located in an external electric field, is also well known:

$$j_{\mathrm{T}}^{\mathrm{sh}} = (1 - R)A_0 T^2 \exp\left(-\frac{\varsigma - W_{\mathrm{sh}}}{k_{\mathrm{B}}T}\right), \tag{2.32}$$

where $W_{\mathrm{sh}} = \sqrt{e^3 E_{\mathrm{F}}}$ is the value of this decrease measured in the electrostatic system of units (ESU); here E_{F} is the strength of the external field.

The field emission is the emission of electrons from a solid under the action of a strong external electric field at a low temperature of the emitter. The phenomenon is based on the tunneling of elecrons through the potential barrier near the body surface. The *tunnel effect* is the transition of a particle through the potential barrier in the case where its total energy is less than the barrier height; this is a pure quantum effect completely impossible in the framework of the classical mechanics. In the field emission, the tunneling becomes possible due to the curvature of the potential barrier (its decrease and "thickening", see Fig. 2.13 in Sect. 2.4.3) under the action of an external field applied to it. The density of field emission current j_{F} (in the emission from a metal into a vaccume) is described by the Fowler–Nordheim formula [18]:

$$j_{\mathrm{F}} = A_{\mathrm{FN}} E_{\mathrm{F}}^2 \exp\left(-\frac{B_{\mathrm{FN}}\,\varsigma^{3/2}}{E_{\mathrm{F}}}\right), \tag{2.33}$$

where A_{FN} and B_{FN} are the Fowler–Nordheim constants,

$$A_{\mathrm{FN}} = \frac{e^3}{8\pi h \varsigma t_{\mathrm{FN}}^2(E_{\mathrm{F}}, \varsigma)}, \quad B_{\mathrm{FN}} = \frac{8\pi\sqrt{2m}}{3he} v_{\mathrm{FN}}(E_{\mathrm{F}}, \varsigma),$$

and $t_{\mathrm{FN}}(E_{\mathrm{F}}, \zeta)$ and $v_{\mathrm{FN}}(E_{\mathrm{F}}, \zeta)$ are some special functions called Nordheim functions [12, 37, 39], which can be represented as functions of a single argument $\hat{y} = \sqrt{e^3 E_{\mathrm{F}}}/\varsigma$:

$$t_{\mathrm{FN}}(\hat{y}) = v_{\mathrm{FN}}(\hat{y}) - \frac{2\hat{y}}{3}\frac{\mathrm{d}v_{\mathrm{FN}}(\hat{y})}{\mathrm{d}\hat{y}}, \tag{2.34}$$

and the explicit form of $v_{\mathrm{FN}}(\hat{y})$ will be given below, see (2.100).

If both factors affecting the emission, i.e., the high temperature and the high strength of the external electric field, are taken into account, then we observe the *thermo-field emission* (or thermo-field electron emission). It is well known [39] that the density of thermo-field emission current j_{TF} is greater than the sum of the densities of the thermo field and field emissions:

$$j_{\mathrm{TF}} > j_{\mathrm{T}} + j_{\mathrm{F}}.$$

The theory of thermo-field emission is described in the vast literature [5, 11, 12, 14, 20, 24, 26, 31, 33, 35, 37]. In this section, we consider some theoretical formulas describing the thermo-field emission.

The density of thermo-field emission current j_{TF} (in the emission from a metal to a vacuum) can be determined by the formula

$$j_{\mathrm{TF}} = e \int_{-\infty}^{\infty} N(W_x) D(W_x) \, dW_x, \qquad (2.35)$$

where W_x is the energy of electrons radiated along the normal to the emitter surface,

$$W_x = \frac{p_x^2}{2m_e} + V(x),$$

and $N(W_x)D(W_x)dW_x$ is the number of electrons emitted from the unit surface per unit time with the energy in the interval of width dW_x. The function $N(W_x)$ is called the support function and the function $D(W_x)$ is the transparency factor of the potential barrier.

2.4.1 Support Function in Metals

As was already mentioned in Sect. 2.1, the electrons inside the emitter obey the Fermi–Dirac distribution (see formula (2.2) and Fig. 2.5 in Sect. 2.1). The density of states per unit volume of the metal is equal to $2/h^3$ [35]. Thus, the number of electrons on the emitter surface per a time interval and an area element is determined by the support function of the form

$$N(W_x) = \frac{2}{h^3} \int_{-\infty}^{\infty} \int_{-\infty}^{\infty} \left(1 + \exp\left(\frac{W_x - W_{\mathrm{F}}}{k_{\mathrm{B}}T} + \frac{p_y^2 + p_z^2}{2mk_{\mathrm{B}}T}\right)\right)^{-1} dp_y dp_z, \qquad (2.36)$$

where the integrand is precisely the function $f_{\mathrm{FD}}(W)$ (see (2.2)). We pass to polar coordinates in expression (2.36):

$$p_y = p_r \cos p_\varphi, \quad p_z = p_r \sin p_\varphi.$$

Then

$$N(W_x) = \frac{2}{h^3} \int_{0}^{\infty} \int_{0}^{2\pi} \left(1 + \exp\left(\frac{W_x - W_{\mathrm{F}}}{k_{\mathrm{B}}T} + \frac{p_r^2}{2mk_{\mathrm{B}}T}\right)\right)^{-1} p_r \, dp_\varphi dp_r. \qquad (2.37)$$

In the obtained expression, the inner integral (with respect to p_φ) is equal to 2π, and in the outer integral, we change the variable

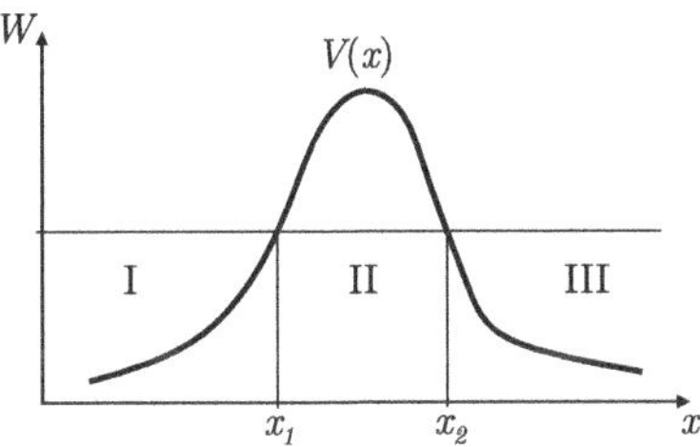

Fig. 2.10 Potential barrier

$$\xi = \frac{p_r^2}{2mk_\mathrm{B}T}.$$

Then expression (2.37) becomes

$$
\begin{aligned}
N(W_x) &= \frac{4\pi m_e k_\mathrm{B}T}{h^3} \int_0^\infty \left(1 + \exp\left(\frac{W_x - W_\mathrm{F}}{k_\mathrm{B}T}\right)e^\xi\right)^{-1} \mathrm{d}\xi \\
&= \frac{4\pi m_e k_\mathrm{B}T}{h^3}\left[\ln\left(\exp\left(\frac{W_x - W_\mathrm{F}}{k_\mathrm{B}T}\right) + 1\right)\right. \\
&\qquad \left. - \ln\left(\exp\left(\frac{W_x - W_\mathrm{F}}{k_\mathrm{B}T}\right)\right)\right].
\end{aligned}
\tag{2.38}
$$

As a result, we obtain the following expression for $N(W_x)$:

$$N(W_x) = \frac{4\pi m_e k_\mathrm{B}T}{h^3}\ln\left(1 + \exp\left(-\frac{W_x - W_\mathrm{F}}{k_\mathrm{B}T}\right)\right). \tag{2.39}$$

2.4.2 Electron Tunneling Through the Potential Barrier

We consider the tunneling of a particle with a certain energy W through an arbitrary potential barrier $V(x)$, see Fig. 2.10. A particle with a certain energy W is incident on the barrier from domain I. Further, the following two versions of the events are possible. In the first version, the particle is reflected from the barrier, in the second, the particle is tunneling through the barrier into domain III. We are interested in the second version.

Certainly, these formulas are widely known but we present them in detail bearing in mind that they are used in numerical calculations later in this book. Additionally, we clarify the notion of the function $v_\mathrm{FN}(y)$ in (2.34).

In Fig. 2.10, the following three domains are distinguished: I is the domain in front of the barrier $x < x_1$ and the particle energy is $W > V(x)$; II is the domain inside the barrier $x_1 < x < x_2$ and $W < V(x)$; III is the domain behind the barrier $x > x_2$ and $W > V(x)$, where x_1 and x_2 are the classical turning points which can be determined from the condition

$$V(x_1) = V(x_2) = W. \tag{2.40}$$

The main value characterizing the tunneling effect is the *transparency factor of the barrier* $D(W)$ which is equal to the absolute value of the ratio of the density of transmitted particles j_{III} to the density of the flow of incident particles j_{I}:

$$D = \left| \frac{j_{\mathrm{III}}}{j_{\mathrm{I}}} \right|, \tag{2.41}$$

where

$$j_k = \frac{\mathrm{i}\hbar}{2m} \left(\frac{\partial u_k^*}{\partial x} u_k - \frac{\partial u_k}{\partial x} u_k^* \right), \quad k = \mathrm{I, III}.$$

Here the superscript $*$ denotes the complex conjugation, and the wave function $u_k = u_k(x)$ is determined by the WKB method[1] from the stationary Schrödinger equation for the motion of a single electron in the direction of the axis x:

$$-\frac{\hbar^2}{2m} \frac{\mathrm{d}^2 u_k}{\mathrm{d}x^2} + V(x)u_k = W u_k, \quad k = \mathrm{I, II, III}. \tag{2.42}$$

Here, as usual, $\hbar$ is the reduced Planck constant, $\hbar = h/2\pi$.

As is known, in the framework of the asymptotic approach in quantum mechanics, the small constant $\hbar$ is taken as a small parameter tending to zero. But the formulas we introduce in asymptotic methods are further used in numerical computations, where the value of the constant $\hbar$ is significant.

Let us determine the solution of Eq. (2.42) in domain I. Here, as was noted above, $x < x_1$ and $W > V(x)$. We seek the solution of Eq. (2.42) in the form

$$u_{\mathrm{I}} = \mathrm{e}^{\frac{\mathrm{i}}{\hbar} S(x)}. \tag{2.43}$$

Then we obtain

$$\frac{\mathrm{d}^2 u_{\mathrm{I}}}{\mathrm{d}x^2} = \left(\frac{\mathrm{i}}{\hbar} S''(x) + \frac{\mathrm{i}^2}{\hbar^2} \left(S'(x) \right)^2 \right) \mathrm{e}^{\frac{\mathrm{i}}{\hbar} S(x)}. \tag{2.44}$$

Substituting (2.43) and (2.44) in Eq. (2.42) and making several transformations, we obtain the equation for the function $S(x)$:

$$\frac{\mathrm{i}}{\hbar} S''(x) + \frac{\mathrm{i}^2}{\hbar^2} \left(S'(x) \right)^2 = \frac{2m}{\hbar^2} (V(x) - W). \tag{2.45}$$

We expand the function $S(x)$ in a power series in $\hbar$:

---

[1] The Wentzel–Kramers–Brillouin method.

$$S(x) = S_0(x) + \frac{\hbar}{i} S_1(x) + \frac{\hbar^2}{i^2} S_2(x) + \cdots , \tag{2.46}$$

substitute expansion (2.46) in Eq. (2.45), and write the coefficients of different powers of $\hbar$:

$$\hbar^{-2} : \quad i^2 \left(S_0'(x) \right)^2 = 2m(V(x) - W), \tag{2.47}$$

$$\hbar^{-1} : \quad i S_0''(x) + 2i S_0'(x) S_1'(x) = 0. \tag{2.48}$$

From (2.47) we obtain the function $S_0(x)$:

$$S_0(x) = \pm \int\limits_{x_1}^{x} \sqrt{2m(W - V(x'))}\, \mathrm{d}x', \tag{2.49}$$

and from (2.48) we obtain the equation for $S_1(x)$:

$$S_1'(x) = -S_0''(x)/2S_0'(x),$$

whence we derive the expression for $S_1(x)$:

$$S_1(x) = -\ln\sqrt{2m(W - V(x))}/2 + \text{const.}$$

As a result, we obtain the following asymptotic solution of Eq. (2.42) in domain I:

$$u_{\mathrm{I}}(x) = \frac{A_1}{\sqrt[4]{2m(W - V(x))}} \exp\left(-\frac{i}{\hbar} \int\limits_{x_1}^{x} \sqrt{2m(W - V(x'))}\, \mathrm{d}x' \right)$$

$$+ \frac{B_1}{\sqrt[4]{2m(W - V(x))}} \exp\left(\frac{i}{\hbar} \int\limits_{x_1}^{x} \sqrt{2m(W - V(x'))}\, \mathrm{d}x' \right). \tag{2.50}$$

In this formula, the term with coefficient B_1 is the wave incident on the barrier, and the term with coefficient A_1 is the wave reflected from the barrier.

We now determine the solution of Eq. (2.42) in domain III. In this domain, everything is as in domain I except for the limits of integration, because $x > x_2$ in this domain. As a result, we obtain the following expression for u_{III}:

$$u_{\mathrm{III}}(x) = \frac{A_3}{\sqrt[4]{2m(W - V(x))}} \exp\left(-\frac{i}{\hbar} \int\limits_{x_2}^{x} \sqrt{2m(W - V(x'))}\, \mathrm{d}x' \right)$$

$$+ \frac{B_3}{\sqrt[4]{2m(W - V(x))}} \exp\left(\frac{i}{\hbar} \int\limits_{x_2}^{x} \sqrt{2m(W - V(x'))}\, \mathrm{d}x' \right). \tag{2.51}$$

Since there is no wave reflected from infinity, we put $A_3 = 0$:

$$u_{\text{III}}(x) = \frac{B_3}{\sqrt[4]{2m(W - V(x))}} \exp\left(\frac{\text{i}}{\hbar} \int_{x_2}^{x} \sqrt{2m(W - V(x'))}\,\text{d}x'\right). \tag{2.52}$$

Let us determine the solution of Eq. (2.42) in domain II. In this domain, everything is as in domain I but it is necessary to take into account that $W < V(x)$ in domain II, which implies that the radicand in $\sqrt{2m(W - V(x))}$ is negative, and hence

$$S_0(x) = \pm\text{i} \int_{x_2}^{x} \sqrt{2m|W - V(x)|}\,\text{d}x';$$

$$S_1(x) = -\ln\left(\sqrt{2m|W - V(x)|}\right)/2 + \text{const.}$$

As a result, we obtain the formula

$$u_{\text{II}}(x) = \frac{A_2}{\sqrt[4]{2m|W - V(x)|}} \exp\left(\frac{1}{\hbar} \int_{x_2}^{x} \sqrt{2m(W - V(x'))}\,\text{d}x'\right)$$

$$+ \frac{B_2}{\sqrt[4]{2m|W - V(x)|}} \exp\left(-\frac{1}{\hbar} \int_{x_2}^{x} \sqrt{2m|W - V(x')|}\,\text{d}x'\right). \tag{2.53}$$

The formulas obtained for u_{I}, u_{II}, and u_{III} hold for all x except for the values in small neighborhoods of the turning points x_1 and x_2. Now we determine the solutions in these neighborhoods.

We consider a small neighborhood of the turning point x_1, where the potential $V(x)$ increases monotonically (see Fig. 2.10), and hence $V'(x_1) > 0$. We expand $V(x)$ in the Taylor series in a neighborhood of the point x_1:

$$V(x) = V(x_1) + V'(x_1) \cdot (x - x_1) + \cdots \tag{2.54}$$

and substitute expansion (2.54) in Eq. (2.42). Since $V(x_1) = W$, we obtain

$$-\frac{\hbar^2}{2m} u''_{x_1} + V'(x_1) \cdot (x - x_1) u_{x_1} = 0. \tag{2.55}$$

We perform the change

$$y = \left(\frac{2m V'(x_1)}{\hbar^2}\right)^{1/3} (x_1 - x) \tag{2.56}$$

and obtain

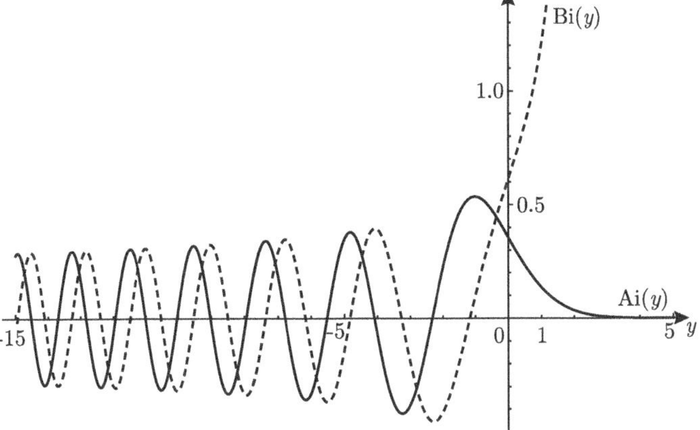

Fig. 2.11 Graphs of the functions Ai (y) and Bi (y)

$$\frac{d^2}{dx^2} = \left(\frac{2mV'(x_1)}{\hbar^2}\right)^{2/3} \frac{d^2}{dy^2}. \tag{2.57}$$

Substituting (2.56) and (2.57) in (2.55), we obtain the equation

$$\frac{d^2 u_{x_1}}{dy^2} + y u_{x_1} = 0. \tag{2.58}$$

It is well known [45] that Eq. (2.58) is called the Airy equation. Its solution is a superposition of the Airy functions Ai (y) and Bi (y):

$$u_{x_1}(y) = \tilde{C}_1 \, \text{Ai}\,(-y) + \tilde{C}_2 \, \text{Bi}\,(-y). \tag{2.59}$$

For real y, the Airy function Ai (y) is determined by the integral

$$\text{Ai}\,(y) = \frac{1}{\pi} \int_0^{\infty} \cos\left(\frac{t^3}{3} + yt\right) dt,$$

and the function Bi (y), by

$$\text{Bi}\,(y) = \frac{1}{\pi} \int_0^{\infty} \left[\exp\left(-\frac{t^3}{3} + yt\right) + \sin\left(\frac{t^3}{3} + yt\right)\right] dt;$$

the graphs of these functions are shown in Fig. 2.11. One can see that, at a certain point, the behavior of these functions varies from oscillating to exponential.

We note that the solution is considered in a small $\hbar$-independent neighborhood of the point x_1, and hence, after the change (2.56), as $\hbar \to 0$, we obtain $y \to -\infty$ for $x > x_1$ and $y \to \infty$ for $x < x_1$. Therefore, we are interested in the behavior of our solution as $y \to \pm\infty$. The asymptotic expressions for the Airy functions for large y are known [45]:

$$\mathrm{Ai}\,(y) = \frac{1}{2\sqrt{\pi}\sqrt[4]{y}}\,\exp\left(-\frac{2}{3}y^{3/2}\right), \quad y \to \infty; \tag{2.60}$$

$$\mathrm{Bi}\,(y) = \frac{1}{\sqrt{\pi}\sqrt[4]{y}}\,\exp\left(\frac{2}{3}y^{3/2}\right), \quad y \to \infty; \tag{2.61}$$

$$\mathrm{Ai}\,(y) = \frac{1}{\sqrt{\pi}\sqrt[4]{-y}}\,\cos\left(\frac{2}{3}(-y)^{3/2} - \frac{\pi}{4}\right), \quad y \to -\infty; \tag{2.62}$$

$$\mathrm{Bi}\,(y) = \frac{1}{\sqrt{\pi}\sqrt[4]{-y}}\,\cos\left(\frac{2}{3}(-y)^{3/2} + \frac{\pi}{4}\right), \quad y \to -\infty. \tag{2.63}$$

Let us return to the change (2.56):

$$y = \frac{(2m\,V'(x_1))^{1/3}}{\hbar^{2/3}}(x_1 - x) = \frac{2m\,V'(x_1)(x_1 - x)}{(2m\,V'(x_1)\hbar)^{2/3}}.$$

It follows from expansion (2.54) that

$$W - V(x) = V'(x_1)(x_1 - x),$$

and hence, we obtain the formula for y:

$$y = \frac{2m(W - V(x))}{(2m\,V'(x_1)\hbar)^{2/3}}. \tag{2.64}$$

The following formula also holds:

$$y^{3/2} = \frac{\sqrt{2m\,V'(x_1)}}{\hbar}(x_1 - x)^{3/2}.$$

We note that

$$\int_{x_1}^{x} \sqrt{x_1 - x'}\,\mathrm{d}x' = -\frac{2}{3}(x_1 - x)^{3/2}.$$

Therefore, we have

$$y^{3/2} = -\frac{3}{2}\frac{1}{\hbar}\int_{x_1}^{x}\sqrt{2m(W - V(x'))}\,dx'.$$

(2.65)

We introduce the notation

$$\gamma_1 = \frac{1}{\sqrt{\pi}}(2m\,V'(x_1)\hbar)^{-1/3};$$

(2.66)

$$\gamma_2 = \frac{1}{\sqrt{\pi}}(2m|V'(x_2)|\hbar)^{-1/3}.$$

(2.67)

Substituting expressions (2.60)–(2.63) in the solution (2.59) and taking into account (2.64)–(2.67) and the fact $y < 0$ for $x > x_1$ and $y > 0$ for $x < x_1$, we finally obtain the wave function in a neighborhood of the point x_1:

$$u_{x_1,\,x<x_1}(x) = \frac{\gamma_1}{\sqrt[4]{2m(W - V)}}\left[\tilde{C}_1\cos\left(\frac{1}{\hbar}\int_{x_1}^{x}\sqrt{2m(W - V(x'))}\,dx' + \frac{\pi}{4}\right)\right.$$
$$\left. + \tilde{C}_2\cos\left(\frac{1}{\hbar}\int_{x_1}^{x}\sqrt{2m(W - V(x'))}\,dx' - \frac{\pi}{4}\right)\right];$$

(2.68)

$$u_{x_1,\,x>x_1}(x) = \frac{\gamma_1}{\sqrt[4]{2m|W - V|}}\left[\frac{\tilde{C}_1}{2}\exp\left(-\frac{1}{\hbar}\int_{x_1}^{x}\sqrt{2m|W - V(x')|}\,dx'\right)\right.$$
$$\left. + \tilde{C}_2\exp\left(\frac{1}{\hbar}\int_{x_1}^{x}\sqrt{2m|W - V(x')|}\,dx'\right)\right].$$

(2.69)

Similarly, we can obtain the solution in a neighborhood of the point x_2, where the potential $V(x)$ decreases monotonically (i.e., $V'(x_2) < 0$). It has the form

$$u_{x_2,\,x<x_2}(x) = \frac{\gamma_2}{\sqrt[4]{2m|W - V|}}\times\left[\frac{\tilde{C}_3}{2}\exp\left(\frac{1}{\hbar}\int_{x_2}^{x}\sqrt{2m|W - V(x')|}\,dx'\right)\right.$$
$$\left. + \tilde{C}_4\exp\left(-\frac{1}{\hbar}\int_{x_2}^{x}\sqrt{2m|W - V(x')|}\,dx'\right)\right];$$

(2.70)

$$u_{x_2,\,x>x_2}(x) = \frac{\gamma_2}{\sqrt[4]{2m(W-V)}}\left[\widetilde{C}_3\cos\left(\frac{1}{\hbar}\int\limits_{x_2}^{x}\sqrt{2m\big(W-V(x')\big)}\,\mathrm{d}x' - \frac{\pi}{4}\right)\right.$$

$$\left. + \widetilde{C}_4\cos\left(\frac{1}{\hbar}\int\limits_{x_2}^{x}\sqrt{2m\big(W-V(x')\big)}\,\mathrm{d}x' + \frac{\pi}{4}\right)\right]. \tag{2.71}$$

Now it only remains to "sew together" the obtained "global" solutions u_{I}, u_{II}, u_{III} with the "local" solutions $u_{x_1,\,x<x_1}$, $u_{x_1,\,x>x_1}$, $u_{x_2,\,x<x_2}$, $u_{x_2,\,x>x_2}$ near the turning points. We begin the "sewing" procedure with domain III, where we obtain the following expression for the wave function (see (2.52)):

$$u_{\mathrm{III}}(x) = \frac{B_3}{\sqrt[4]{2m(W-V(x))}}\exp\left(\frac{\mathrm{i}}{\hbar}\int\limits_{x_2}^{x}\sqrt{2m(W-V(x'))}\,\mathrm{d}x'\right),$$

and in the right-hand neighborhood of the point x_2, we obtain expression (2.71):

$$u_{x_2,\,x>x_2}(x) = \frac{\gamma_2}{\sqrt[4]{2m(W-V)}}\left[\widetilde{C}_3\cos\left(\frac{1}{\hbar}\int\limits_{x_2}^{x}\sqrt{2m\big(W-V(x')\big)}\,\mathrm{d}x' - \frac{\pi}{4}\right)\right.$$

$$\left. + \widetilde{C}_4\cos\left(\frac{1}{\hbar}\int\limits_{x_2}^{x}\sqrt{2m\big(W-V(x')\big)}\,\mathrm{d}x' + \frac{\pi}{4}\right)\right].$$

Therefore, by setting

$$\widetilde{C}_4 = \mathrm{i}\widetilde{C}_3 \tag{2.72}$$

in (2.71) and taking into account that $\cos\left(x - \frac{\pi}{2}\right) = \sin(x)$, we obtain

$$u_{x_2,\,x>x_2} = \frac{\gamma_2}{\sqrt[4]{2m(W-V)}}\widetilde{C}_3\exp\left(\frac{1}{\hbar}\int\limits_{x_2}^{x}\sqrt{2m\big(W-V(x')\big)}\,\mathrm{d}x' - \frac{\pi}{4}\right). \tag{2.73}$$

Comparing expressions (2.52) and (2.73), we see that

$$B_3 = \gamma_2\widetilde{C}_3\exp\left(-\mathrm{i}\frac{\pi}{4}\right). \tag{2.74}$$

In domain II, we have the solution (see (2.53)):

$$u_{\text{II}}(x) = \frac{A_2}{\sqrt[4]{2m|W - V(x)|}} \exp\left(\frac{1}{\hbar} \int_{x_2}^{x} \sqrt{2m(W - V(x'))}\, dx'\right)$$

$$+ \frac{B_2}{\sqrt[4]{2m|W - V(x)|}} \exp\left(-\frac{1}{\hbar} \int_{x_2}^{x} \sqrt{2m|W - V(x')|}\, dx'\right),$$

and the solution to the left of the turning point x_2 (see (2.70)) with regard to (2.72) has the form

$$u_{x_2,\, x < x_2}(x) = \frac{\gamma_2}{\sqrt[4]{2m|W - V|}} \tilde{C}_3 \left[\frac{1}{2} \exp\left(\frac{1}{\hbar} \int_{x_2}^{x} \sqrt{2m|W - V(x')|}\, dx'\right) \right.$$

$$\left. + \mathrm{i}\exp\left(-\frac{1}{\hbar} \int_{x_2}^{x} \sqrt{2m|W - V(x')|}\, dx'\right) \right]. \tag{2.75}$$

Therefore, comparing (2.75) with (2.53), we obtain

$$A_2 = \gamma_2 \tilde{C}_3 / 2, \quad B_2 = \mathrm{i}\gamma_2 \tilde{C}_3,$$

and taking (2.74) into account, we see that

$$A_2 = \frac{B_3}{2} \exp\left(\mathrm{i}\frac{\pi}{4}\right); \quad B_2 = \mathrm{i}B_3 \exp\left(\mathrm{i}\frac{\pi}{4}\right). \tag{2.76}$$

Relations (2.76) must also be satisfied on the right of the first turning point (see (2.69)):

$$u_{x_1,\, x > x_1}(x) = \frac{\gamma_1}{\sqrt[4]{2m|W - V|}} \left[\frac{\tilde{C}_1}{2} \exp\left(-\frac{1}{\hbar} \int_{x_1}^{x} \sqrt{2m|W - V(x')|}\, dx'\right) \right.$$

$$\left. + \tilde{C}_2 \exp\left(\frac{1}{\hbar} \int_{x_1}^{x} \sqrt{2m|W - V(x')|}\, dx'\right) \right].$$

We transform the integral in the exponents as

$$\int_{x_1}^{x} \sqrt{2m|W - V(x')|}\, dx' = \int_{x_1}^{x_2} \sqrt{2m|W - V(x')|}\, dx' + \int_{x_2}^{x} \sqrt{2m|W - V(x')|}\, dx'$$

and introduce the notation

$$H \stackrel{\text{def}}{=} \int_{x_1}^{x_2} \sqrt{2m\left|W - V(x')\right|}\, dx'. \tag{2.77}$$

Then, comparing the expression

$$u_{x_1,\, x > x_1}(x) = \frac{\gamma_1}{\sqrt[4]{2m|W - V|}} \left[\frac{\tilde{C}_1}{2} \exp\left(-\frac{1}{\hbar} \int_{x_2}^{x} \sqrt{2m\left|W - V(x')\right|}\, dx' \right) e^{-H/\hbar} \right.$$

$$\left. + \tilde{C}_2 \exp\left(\frac{1}{\hbar} \int_{x_2}^{x} \sqrt{2m\left|W - V(x')\right|}\, dx' \right) e^{H/\hbar} \right]$$

with (2.53), we obtain

$$\tilde{C}_1 = \frac{2B_2}{\gamma_1} \exp\left(\frac{H}{\hbar} \right); \quad \tilde{C}_2 = \frac{A_2}{\gamma_1} \exp\left(-\frac{H}{\hbar} \right).$$

Therefore, with regard to (2.76), we have

$$\tilde{C}_1 = 2i\frac{B_3}{\gamma_1} \exp\left(\frac{H}{\hbar} + i\frac{\pi}{4} \right); \quad \tilde{C}_2 = \frac{B_3}{2\gamma_1} \exp\left(-\frac{H}{\hbar} + i\frac{\pi}{4} \right). \tag{2.78}$$

Now we consider domain I. In this domain, we obtain solution (2.50) of Schrödinger equation (2.42) which has the form

$$u_1(x) = \frac{A_1}{\sqrt[4]{2m(W - V(x))}} \exp\left(-\frac{i}{\hbar} \int_{x_1}^{x} \sqrt{2m(W - V(x'))}\, dx' \right)$$

$$+ \frac{B_1}{\sqrt[4]{2m(W - V(x))}} \exp\left(\frac{i}{\hbar} \int_{x_1}^{x} \sqrt{2m(W - V(x'))}\, dx' \right),$$

and after replacement of the cosines with exponentials by the formula $\cos(x) = (e^{ix} + e^{-ix})/2$, the solution to the left of the turning point x_1 (see (2.68)) becomes

$$u_{x_1,\, x < x_1}(x) = \frac{\gamma_1}{\sqrt[4]{2m(W - V)}} \frac{1}{2} \tilde{C}_1 \exp\left(\frac{i}{\hbar} \int_{x_1}^{x} \sqrt{2m(W - V(x'))}\, dx' + i\frac{\pi}{4} \right)$$

$$+ \tilde{C}_1 \exp\left(\frac{-i}{\hbar} \int_{x_1}^{x} \sqrt{2m(W - V(x'))}\, dx' - i\frac{\pi}{4} \right)$$

$$+ \widetilde{C}_2 \exp\left(\frac{i}{\hbar} \int_{x_1}^{x} \sqrt{2m\left(W - V(x')\right)}\, dx' - i\frac{\pi}{4} \right)$$

$$+ \widetilde{C}_2 \exp\left(\frac{-i}{\hbar} \int_{x_1}^{x} \sqrt{2m\left(W - V(x')\right)}\, dx' + i\frac{\pi}{4} \right). \tag{2.79}$$

Comparing (2.79) with (2.50), we obtain

$$A_1 = \frac{\gamma_1}{2}\left(\widetilde{C}_1 e^{-i\pi/4} + \widetilde{C}_2 e^{i\pi/4} \right); \quad B_1 = \frac{\gamma_1}{2}\left(\widetilde{C}_1 e^{i\pi/4} + \widetilde{C}_2 e^{-i\pi/4} \right). \tag{2.80}$$

Taking (2.78) and the relation $e^{i\pi/2} = i$ into account, from (2.80) we finally derive

$$A_1 = iB_3\left(e^{H/\hbar} + \frac{1}{4} e^{-H/\hbar} \right); \quad B_1 = B_3\left(-e^{H/\hbar} + \frac{1}{4} e^{-H/\hbar} \right). \tag{2.81}$$

Thus, we expressed all unknown coefficients of the obtained solutions of Schrödinger equation (2.42) in terms of the coefficient B_3, and we need not determine the explicit form of this coefficient (see below). The solution "sewing" procedure is complete.

Now we determine the density of the transmitted particles j_{III}:

$$j_{\mathrm{III}} = \frac{i\hbar}{2m}\left(\frac{\partial u_{\mathrm{III}}^*}{\partial x} u_{\mathrm{III}} - \frac{\partial u_{\mathrm{III}}}{\partial x} u_{\mathrm{III}}^* \right).$$

We introduce the notation $p(x) = \sqrt{2m(W - V)}$. Then

$$u_{\mathrm{III}}(x) = \frac{B_3}{\sqrt{p(x)}} \exp\left(\frac{i}{\hbar} \int_{x_2}^{x} p(x')\, dx' \right),$$

and the density of transmitted particles becomes

$$j_{\mathrm{III}} = \frac{i\hbar}{2m}\left[-B_3^* \frac{1}{2} p^{-3/2} p' \exp\left(-\frac{i}{\hbar} \int_{x_2}^{x} p\, dx' \right) \right.$$

$$\times\, B_3 p^{-1/2} \exp\left(\frac{i}{\hbar} \int_{x_2}^{x} p\, dx' \right) - B_3^* p^{-1/2} \frac{i}{\hbar} p \exp\left(-\frac{i}{\hbar} \int_{x_2}^{x} p\, dx' \right)$$

$$\times\, B_3 p^{-1/2} \exp\left(\frac{i}{\hbar} \int_{x_2}^{x} p\, dx' \right) + B_3 \frac{1}{2} p^{-3/2} p' \exp\left(\frac{i}{\hbar} \int_{x_2}^{x} p\, dx' \right)$$

$$\times B_3^* p^{-1/2} \exp\left(-\frac{i}{\hbar}\int_{x_2}^{x} p\,dx'\right) - B_3 p^{-1/2}\frac{i}{\hbar}p\exp\left(\frac{i}{\hbar}\int_{x_2}^{x} p\,dx'\right)$$

$$\times B_3^* p^{-1/2}\exp\left(-\frac{i}{\hbar}\int_{x_2}^{x} p\,dx'\right)\Bigg] = \frac{i\hbar}{2m}|B_3|^2\left(-2\frac{i}{\hbar}\right). \tag{2.82}$$

Therefore, we have

$$j_{\text{III}} = |B_3|^2/m. \tag{2.83}$$

Now we calculate the density of incident particles j_{I}:

$$j_{\text{I}} = \frac{i\hbar}{2m}\left(\frac{\partial \tilde{u}_{\text{I}}^*}{\partial x}\tilde{u}_{\text{I}} - \frac{\partial \tilde{u}_{\text{I}}}{\partial x}\tilde{u}_{\text{I}}^*\right),$$

where $\tilde{u}_{\text{I}}$ denotes the wave function component incident on the barrier u_{I}:

$$\tilde{u}_{\text{I}} = \frac{B_1}{\sqrt{p(x)}}\exp\left(\frac{i}{\hbar}\int_{x_1}^{x} p(x')\,dx'\right).$$

Therefore, by similar transformations (see (2.82)), we have

$$j_{\text{I}} = |B_1|^2/m. \tag{2.84}$$

Substituting obtained expressions (2.83) and (2.84) in (2.41), we obtain the expression for D:

$$D = \left|\frac{j_{\text{III}}}{j_{\text{I}}}\right| = \frac{|B_3|^2}{|B_1|^2}. \tag{2.85}$$

We substitute the expression for the coefficient B_1 from formula (2.81) in formula (2.85) and obtain

$$D = \left(-e^{H/\hbar} + \frac{1}{4}e^{-H/\hbar}\right)^{-2} = e^{-2H/\hbar}\left(1 - \frac{1}{4}e^{-2H/\hbar}\right)^{-2}. \tag{2.86}$$

We note that $e^{-2H/\hbar} \ll 1$ as $\hbar \to 0$. Therefore, we have the formula for D:

$$D \approx \exp\left(-2\frac{H}{\hbar}\right). \tag{2.87}$$

Substituting expression (2.77) in (2.87), we obtain the final expression for $D(W)$:

Fig. 2.12 Electron and its image

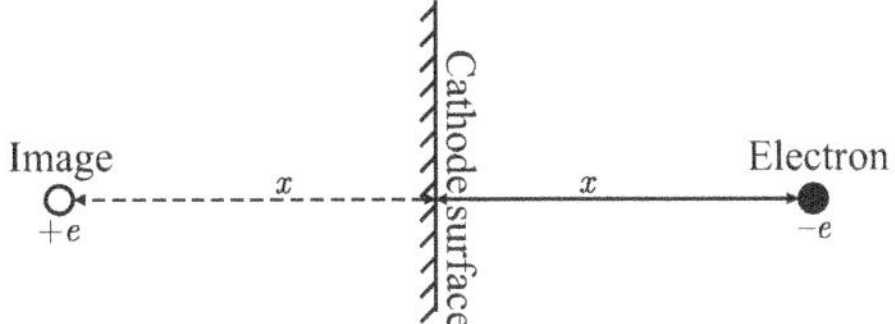

$$D(W) \approx \exp\left(-\int_{x_1}^{x_2} \sqrt{\frac{8m}{\hbar^2}(V(x) - W)}\,\mathrm{d}x \right). \tag{2.88}$$

In general, in addition to formula (2.87), there exist other formulas for the barrier transparency factor obtained by other methods. The formula $D(W) \approx P \exp(-2H/\hbar)$, where P is a correction coefficient, is given in [32], $D(W) \approx P \exp(-2H/\hbar)/[1 + P \exp(-2H/\hbar)]$ in [19], and $D(W) \approx 1/[1 + P \exp(2H/\hbar)]$ in [29]. These formulas were tested in the paper [16], where it was shown that formula (2.87) works well only in the case of an ideally smooth barrier. But to calculate the correction coefficient P is a rather difficult problem, and we do not consider it here but use formula (2.88) in what follows.

2.4.3 Formula for the Barrier Transparency Factor in the Case of Field Emission Cathode

Now we use our model to describe the tunnel effect. First, we consider the case without external electric field, where the potential barrier has a very simple structure. An electron moving in a vacuum at a certain distance x from the cathode surface induces a distributed positive charge on the cathode surface [10, 12]. The field arising in this case between the electron and the induced charge is equivalent to the field between the electron and the effective positive charge equal to the electron charge and is located at the point of mirror image of the electron, which is called the *image charge* (see Fig. 2.12).

Therefore, in the ESU system, the force maintaining the electron is equal to

$$F_0(x) = -\frac{e^2}{(2x)^2}. \tag{2.89}$$

The potential $V_0(x)$ has the form (also see Fig. 2.13a):

$$V_0(x) = -\int_x^\infty \frac{e^2}{(2x)^2}\,\mathrm{d}x = -\frac{e^2}{4x}. \tag{2.90}$$

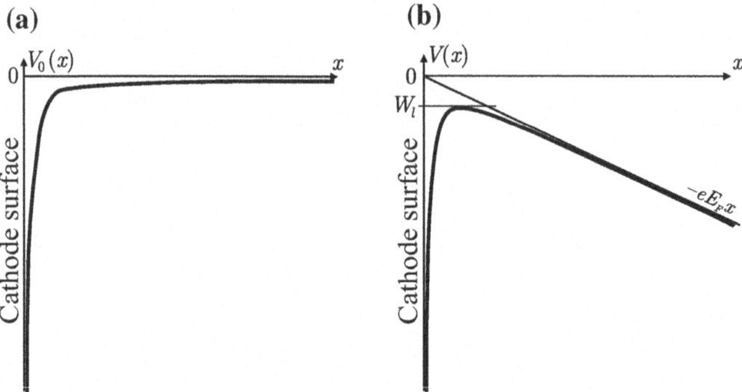

Fig. 2.13 Potential barrier on the cathode–vacuum interface: **a** no external field, **b** an external field is applied

We note that $\lim_{x \to 0} V_0(x) = -\infty$. But formula (2.90) for $V_0(x)$ is of course false for x of the order of interatomic distances. In general, formula (2.89) for the force $F_0(x)$ holds at distances $x \geqslant a$, where a is approximately equal to the crystal lattice constant (and this is a very small quantity, for example, it is approximately equal to 5.43×10^{-10} m for silicon). At a distance $0 < x < a$, the influence of the cathode atomic structure is still large, but at this distance, one can assume that the force is approximately equal to a certain constant [12]. Therefore, for $0 < x < a$, the potential is equal to a constant,[2] and for small positive $x \geqslant a$, the potential is determined by formula (2.90).

The application of an external electric field results in a curvature of the potential barrier, see Fig. 2.13b. The potential of the external electric field E_F is equal to $-eE_Fx$. Therefore, the potential barrier has the form

$$V(x) = -\frac{e^2}{4x} - eE_Fx, \quad x \geqslant a. \tag{2.91}$$

Recall that we are interested in the domain $x \geqslant a$, i.e., in the domain containing emitted electrons (the electrons that moved away from the emitter surface at a distance at which the cathode atomic structure ceases to affect them). In Fig. 2.13b, one can see that the potential becomes lower and narrower as the strength of the applied external electric field increases, and therefore, the electron emission becomes possible. The height of the barrier decreases as the external electric field increases, and this effect is called the *Schottky effect* [37]. The value of the variation in the barrier height is equal to $|W_l|$ (see (2.93)).

Let us return to the expression for the barrier transparency factor. We further assume that formula (2.88) is "exact":

---

[2]This quantity does not play any role in studying the problem of field emission.

$$D(W) = \exp\left(-\int_{x_1}^{x_2} \sqrt{\frac{8m}{\hbar^2}(V(x) - W)}\,dx\right).$$

(2.92)

This formula is correct only of the energies W that do not exceed the value

$$W_l = -\sqrt{e^3 E_{\mathrm{F}}}$$

(2.93)

equal to the maximal value of $V(x)$. If $W > W_l$, then the barrier transparency is $D(W) = 1$ (i.e., there is no barrier, see Fig. 2.13b). Substituting the expression for $V(x)$ from (2.91) in formula (2.92), we obtain

$$-\ln(D) = \int_{x_1}^{x_2} \sqrt{\frac{8m}{\hbar^2}\left(-eE_{\mathrm{F}}x - \frac{e^2}{4x} + |W|\right)}\,dx,$$

(2.94)

where the coordinates of the turning point are

$$x_1 = \frac{|W|}{2eE_{\mathrm{F}}}\left(1 - \sqrt{1 - \frac{e^3 E_{\mathrm{F}}}{W^2}}\right); \quad x_2 = \frac{|W|}{2eE_{\mathrm{F}}}\left(1 + \sqrt{1 - \frac{e^3 E_{\mathrm{F}}}{W^2}}\right).$$

We perform the change

$$y = \frac{\sqrt{e^3 E_{\mathrm{F}}}}{|W|}$$

(2.95)

and obtain the new variable of integration

$$\xi = \frac{2eE_{\mathrm{F}}}{|W|}x.$$

(2.96)

As a result, after transformations (2.95) and (2.96), formula (2.94) becomes

$$-\ln(D) = \frac{\sqrt{m|W|^3}}{\hbar e E_{\mathrm{F}}} \int_{1-\sqrt{1-y^2}}^{1+\sqrt{1-y^2}} \sqrt{-\xi^2 + 2\xi - y^2}\,\frac{d\xi}{\sqrt{\xi}}.$$

(2.97)

The substitution $\eta = \sqrt{\xi}$ takes (2.97) to the form

$$-\ln(D) = \frac{2\sqrt{m|W|^3}}{\hbar e E_{\mathrm{F}}} \int_{\tilde{b}}^{\tilde{a}} \sqrt{(a^2 - \eta^2)(\eta^2 - b^2)}\,d\eta,$$

(2.98)

where $a = \sqrt{1 + \sqrt{1 - y^2}}$ and $b = \sqrt{1 - \sqrt{1 - y^2}}$.

Simplifying the integral in the right-hand side of (2.98), we obtain the final formula for the barrier transparency factor

$$D(W) = \exp\left(-\frac{4\sqrt{2m|W|^3}}{3\hbar e E_{\mathrm{F}}} v_{\mathrm{FN}}(y)\right), \tag{2.99}$$

where

$$v_{\mathrm{FN}}(y) = \begin{cases} \sqrt{y+1}\left[\mathrm{E}\left(\dfrac{1-y}{1+y}\right) - y\mathrm{K}\left(\dfrac{1-y}{1+y}\right)\right], & y \leqslant 1; \\[2ex] -\sqrt{\dfrac{y}{2}}\left[(y+1)\mathrm{K}\left(\dfrac{y-1}{2y}\right) - 2\mathrm{E}\left(\dfrac{y-1}{2y}\right)\right], & y > 1; \end{cases} \tag{2.100}$$

here $\mathrm{E}(k)$ and $\mathrm{K}(k)$ are elliptic integrals of the first and second kind, respectively [1]. The obtained expression is true if x_1 and x_2 are real numbers. For this, it is necessary that $W \leqslant W_l$, which is true in our case.

Our model only requires the values of the function $v_{\mathrm{FN}}(y)$ for $y \leqslant 1$, because $W \in (-\infty, W_l]$, where $W_l < 1$, and the variable $y \in (0, |W_l|)$ by the change (2.95). To simplify the calculations, instead of the function $v_{\mathrm{FN}}(y)$, $y \leqslant 1$, defined by (2.100), we construct its approximation by a polynomial of degree two which has the form

$$v_{\mathrm{apr}}(y) = 0.95 - 1.03y^2. \tag{2.101}$$

In Fig. 2.14, the function $v_{\mathrm{FN}}(y)$, $y \leqslant 1$, is compared with its approximation $v_{\mathrm{apr}}(y)$. We note that this approximation of the function $v_{\mathrm{FN}}(y)$ is not unique; there are other approximations (see [15] for details).

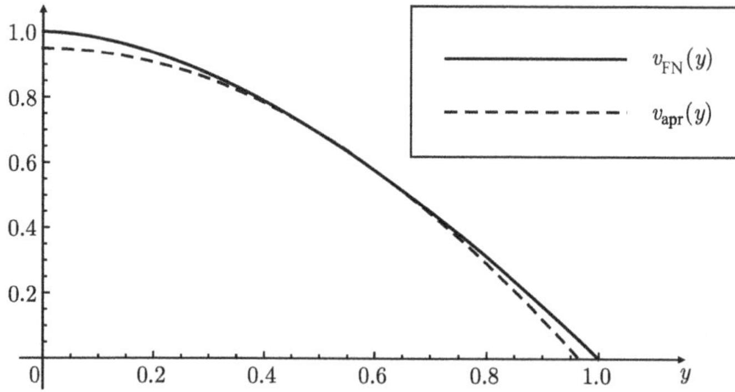

Fig. 2.14 Comparison of $v_{\mathrm{FN}}(y)$, $y \leqslant 1$, with its approximation v_{apr}

2.4.4 Emission Current Density in Metals

Despite the fact that formula (2.88) is derived by approximate methods, we will treat it as an "exact" formula. Thus, we do not consider the situations where, for example, the WKB-asymptotics cannot be used (for fields of very high strength ($\approx 10^{10}$ V/m), when the potential barrier on the emitter surface becomes very narrow).

We substitute obtained formulas (2.39) and (2.88) in the formula for the current density (2.35). As a result, under the assumption that $D(W_x) = 1$ for $W_x > W_l$, we obtain the following formula for the thermo-field emission density

$$
j_{\mathrm{TF}}(T, E_\mathrm{F}, W_F) = A_{\mathrm{TF}} \Bigg[\int\limits_{-\infty}^{W_l} \ln\left(1 + \exp\left(-\frac{W_x - W_\mathrm{F}}{k_\mathrm{B}T}\right)\right)
$$
$$
\times \exp\left(-\frac{4\sqrt{2m|W_x|^3}}{3\hbar e E_\mathrm{F}} v_{\mathrm{FN}}(y)\right) \mathrm{d}W_x
$$
$$
+ \int\limits_{W_l}^{\infty} \ln\left(1 + \exp\left(-\frac{W_x - W_\mathrm{F}}{k_\mathrm{B}T}\right)\right) \mathrm{d}W_x \Bigg], \qquad (2.102)
$$

where

$$
A_{\mathrm{TF}} = \frac{4\pi m_e k_\mathrm{B} T e}{h^3}; \quad W_l = -\sqrt{e^3 E_\mathrm{F}}; \quad y = \frac{\sqrt{e^3 E_\mathrm{F}}}{|W_x|},
$$

and the function $v_{\mathrm{FN}}(y)$ was defined above, see (2.100), (2.101).

2.4.5 Specific Characteristics of Field Emission from a Semiconductor Cathode

Now we pass to the description of field emission from a semiconductor cathode, where there arise an effect that cannot arise in metals [10, 12, 43]. First, in the semiconductor, it is necessary to consider the electron emission not only from the conduction band but also from the valence band separated from the conduction band by the band gap that does not contain electrons. Second, it is necessary to take into account that the external electric field penetrates inside the semiconductor which "bends" the energy levels.

The derivation of the formula for the current density due to the emission from the valence band in semiconductors is in the whole similar to the case of metals, but for the valence band, the distribution of electrons over the energies $N(W)D(W)$ has no maximum near the Fermi level which lies already inside the band gap. The maximal energy level which can be occupied by an electron in the valence band is equal to $-(\psi + W_g)$, where ψ is the width of the conduction band and W_g is the width of the band gap (see (2.1)). The transparency factor of the potential barrier $D(W)$ has maximum at $W = -W^\psi$, where

$$W^\psi = \psi + W_g,$$

and exponentially decreases with W.

Another difference between the emission from the valence band of a semiconductor and the emission from a metal is the value of the potential of the electrostatic image of the emitted electron. Because of the dielectric properties of semiconductors, this potential becomes [43]

$$V_0^s(x) = -\frac{\varepsilon_s - 1}{\varepsilon_s + 1}\frac{e^2}{4x}, \tag{2.103}$$

where ε_s is the permittivity of the semiconductor. Then the quantities y (see (2.95)) and W_l (see (2.93)) become

$$y^s = \sqrt{\frac{\varepsilon_s - 1}{\varepsilon_s + 1}}\frac{\sqrt{e^3 E_F}}{|W|}; \quad W_l^s = -\frac{\varepsilon_s - 1}{\varepsilon_s + 1}\sqrt{e^3 E_F}.$$

Recall that we consider the field emission from the silicon cathode. The permittivity of silicon ε_s is approximately equal to 11.9 [10], and hence,

$$\frac{\varepsilon_s - 1}{\varepsilon_s + 1} \approx 0.92.$$

The formula for the density of the current due to emission from the conductivity band is similar to the formula for metals with the only difference that, in the semiconductor, the minimal energy of electrons is determined by the bottom of the conduction band, and hence the lower limit of the integral is equal to $-\psi$.

We write the formula for the current density with regard to the distribution of electrons over the bands in the semiconductor:

$$
\begin{aligned}
j_{\text{em}}^0 = A_{\text{TF}}\Bigg[& \int_{-\infty}^{-W^\psi} \ln\left(1 + e^{(W_F - W)/k_B T}\right) \exp\left(-\frac{4\sqrt{2m|W|^3}}{3\hbar e E_F} v_{\text{FN}}(y^s)\right) dW \\
& + \int_{-\psi}^{W_l^s} \ln\left(1 + e^{(W_F - W)/k_B T}\right) \exp\left(-\frac{4\sqrt{2m|W|^3}}{3\hbar e E_F} v_{\text{FN}}(y^s)\right) dW \\
& + \int_{W_l^s}^{\infty} \ln\left(1 + e^{(W_F - W)/k_B T}\right) dW \Bigg],
\end{aligned}
$$

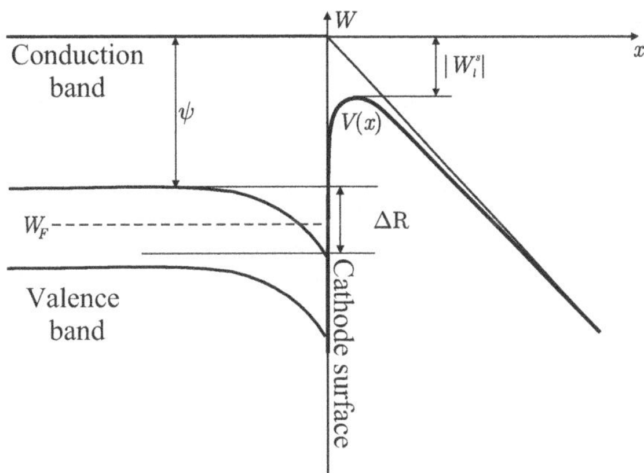

Fig. 2.15 Bending of energy bands in the semiconductor under the action of an external electric field

where the constant A_{TF} and the function $v_{FN}(y)$ were determined previously, see (2.102).

Now we consider the second effect observed in semiconductors, namely, the field penetration into the interior of the semiconductor [12, 43]. In metals, the field penetration depth is negligibly small due to a very large concentration of free electrons. But in semiconductors, the penetration depth is approximately proportional to the square root of the concentration of free carriers [12].

For example, in semiconductors, if the concentration of free carriers is $\sim 10^{16}\,\mathrm{cm}^{-3}$, then the penetration depth is $\sim 10^{-6}\,\mathrm{cm}$, which is approximately equal to 500 atomic layers, and in several cases, this depth is even greater. In this connection, the concentration of free carriers near the surface can significantly differ from the concentrations at the depth, i.e., the external electric field "bends" the conduction band and the valence band downwards near the surface as is shown in Fig. 2.15.

As is known [10], the "bending" of a band near the semiconductor surface ΔR is approximately equal to $\nu E_F^{4/5}$, where $\nu = 4.5 \times 10^{-7}\varepsilon^{-2/5}$. We modify the formula for the emission current density in a semiconductor with this effect taken into account. As a result, we obtain

$$
\begin{aligned}
j_{em} = A_{TF}\Bigg[&\int_{-\infty}^{-W_1} \ln\left(1 + e^{(W_F - W)/k_B T}\right) \exp\left(-\frac{4\sqrt{2m|W|^3}}{3\hbar e E_F} v_{FN}(y^s)\right) dW \\
&+ \int_{-W_2}^{W_i^s} \ln\left(1 + e^{(W_F - W)/k_B T}\right) \exp\left(-\frac{4\sqrt{2m|W|^3}}{3\hbar e E_F} v_{FN}(y^s)\right) dW \\
&+ \int_{W_i^s}^{\infty} \ln\left(1 + e^{(W_F - W)/k_B T}\right) dW \Bigg],
\end{aligned}
\tag{2.104}
$$

where $W_1 = \psi + \nu E_F^{4/5} + W_g$, $W_2 = \psi + \nu E_F^{4/5}$, and the constant A_{TF} and the function $\nu_{\mathrm{FN}}(y)$ were determined previously, see (2.102). We also note that there are other effects typical of semiconductors, for example, the existence of surface energy states of electrons [12]. The field generated by negative charges near the surface can also penetrate into the depth of the conductor. In this case, there is a deficit of electrons near the semiconductor surface and the energy bands "bend" upwards (i.e., oppositely to the effect when the bands "bend" downwards) thus forming an additional "internal" potential barrier. To calculate the value of this effect is a very difficult problem, and hence we do not consider this effect here.

2.4.6 *Approximation of the Formula for the Emission Current Density*

Now we consider an approximate formula for the emission current density. In [39], an approximation of the formula for the emission current density (2.4.4) is given, which has the form (in SI units)

$$j_{\mathrm{FN}}(T, E_{\mathrm{F}}, \varsigma) = \frac{e^3 E_{\mathrm{F}}^2}{8\pi h \varsigma t^2(\hat{y}_s)} \exp\left(\frac{-4\sqrt{2 m_e \varsigma^3}}{3 \hbar e E_{\mathrm{F}}} \nu_{\mathrm{FN}}(\hat{y}_s)\right)$$

$$\times \left[\frac{\pi k_{\mathrm{B}} T}{d_{\mathrm{FN}}(E_{\mathrm{F}}, \varsigma, \hat{y}_s)}\right]\left[\sin\left(\frac{\pi k_{\mathrm{B}} T}{d_{\mathrm{FN}}(E_{\mathrm{F}}, \varsigma, \hat{y}_s)}\right)\right]^{-1}, \quad (2.105)$$

where

$$\hat{y}_s = \sqrt{\frac{e^3 E_{\mathrm{F}}}{4\pi \varepsilon_0}\frac{1}{\varsigma}}; \quad (2.106)$$

$$d_{\mathrm{FN}}(E_{\mathrm{F}}, \varsigma, \hat{y}_s) = \frac{\hbar e E_{\mathrm{F}}}{2\sqrt{2 m_e \varsigma} t_{\mathrm{FN}}(\hat{y}_s)}; \quad (2.107)$$

E_{F} is the strength of the external electric field, ς is the work function (in metals, it is equal to $\varsigma = -W_{\mathrm{F}}$, where W_{F} is the energy of the Fermi level), and T is the temperature. The process of determining this approximation is described in [14, 37]. Here the functions $t_{\mathrm{FN}}(y)$ and $\nu_{\mathrm{FN}}(y)$ were determined previously, see (2.34) and (2.100). Formula (2.105) for $T = 0$ is known as the Fowler–Nordheim formula (see (2.33)), and this formula is sometimes used under the assumption that $t_{\mathrm{FN}}(\hat{y}_s) = 1$ and $\nu_{\mathrm{FN}}(\hat{y}_s) = 1$, which allows one to avoid the calculations of elliptic integrals. But in this simplified form, approximation (2.105) is incorrect for some values of the amplitude for all values of the electric field intensity and the work function.

More precisely, despite the fact that $t_{\mathrm{FN}}(\hat{y}_s) \approx 1$ and $\nu_{\mathrm{FN}}(\hat{y}_s) \approx 1$ for weak fields, the absolute value of the coefficient

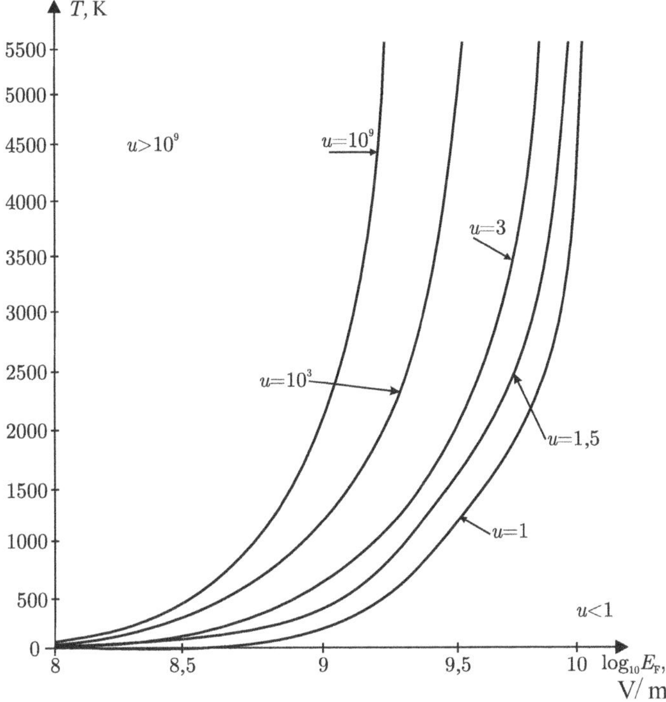

Fig. 2.16 Lines of constant ratio $u = j_{TF}/j_{FN}$ for $\varsigma = 3.5\,\mathrm{eV}$ given in [39]

$$\frac{-4\sqrt{2m_e\varsigma^3}}{3\hbar e E_F}$$

of the function $v_{FN}(\hat{y})$ in the exponent in formula (2.105) is very large. This implies that even small deviations $v_{FN}(\hat{y}_s)$ from 1 lead to a significant error in calculations of the current density j_{FN}. For the values of the electric field strength encountered in practice, the current density values obtained by such an approximation can differ by 10^2–10^6 times [39], which, of course, is unacceptable.

In Fig. 2.16 one can see that the possibility of using the Fowler–Nordheim approximation (2.105) is restricted by small values of the temperature. On the other hand, in the case of strong fields, expression (2.105) also does not work at small temperatures because of zeros of the sine function.

To obtain more precise and simultaneously simple approximate formulas for the thermo-field emission current, one can use the following approximation of the function $v_{FN}(y)$ [39]:

$$\tilde{v}_a(y) = 1 - y^{\beta_y}. \tag{2.108}$$

In this formula, β_y is a numerical parameter which can be optimized to obtain the minimal error in the domain $\hat{y}_s \in [0, 1]$ we are interested in, and its value will be given below in Sect. 2.4.8. A more precise approximation of $v_{FN}(y)$ with a comparatively small error is given in [25]. One more approximation of the function $v_{FN}(y)$ was previously given in (2.101).

The following approximation of the function $t_{FN}(y)$ was proposed in [39]:

$$\widetilde{t}_a(y) = 1 - \alpha y^{\beta}. \tag{2.109}$$

This formula gives a good approximation for two distinct sets of parameters α and β: (α_1, β_1) and (α_2, β_2), because the function $t_{FN}(y)$ varies weakly. The numerical values of the parameters α_1, α_2, β_1, and β_2, as well as the mean and maximal approximation errors arising when using them, are given in Sect. 2.4.8.

Formulas (2.108), (2.109) can be used to obtain an acceptable approximation of Fowler–Nordheim formula (2.105). But one can also use more precise approximation methods in which the parameters are optimized according to the "exact" density of the current given by formula (2.102) [39]:

$$j_{Fit}(E_F, \varsigma) = \frac{q_1 E_F^{q_2}}{\varsigma^{q_3}} \left(1 + \frac{q_4 \varsigma^{q_5}}{E_F^{q_6}} \right) \exp \left(-\frac{q_7 \varsigma^{q_8}}{E_F^{q_9}} + q_{10} \varsigma^{q_{11}} E_F^{q_{12}} \right), \tag{2.110}$$

where the parameters $q_1, \ldots, q_{12}$ are given in Sect. 2.4.8.

We note that, in general, it is difficult to obtain approximate formulas for (2.102) (for metals) and (2.104) (for semiconductors), and one has to calculate the emission current density directly by formula (2.104), which we precisely do in the numerical simulation whose results are presented in Chap. 4.

2.4.7 Nottingham Effect

An important role in the theory of thermo-field emission is played by the Nottingham effect, i.e., the phenomenon of the cathode cooling (heating) under the condition that the average energy of emitted electrons $\mathcal{E}_{em}$ is above (below) the Fermi level W_F of the cathode material. This effect was discovered by Nottingham in 1941 [38] and has actively been studied since that time by several authors, for example, in [4, 6–9, 13, 34, 39]. The Nottingham effect due to the difference $\mathcal{E}_{Nott}$ between the average energy of electrons leaving the cathode $\mathcal{E}_{em}$, and the average energy of electrons approaching the emitter surface from the electric circuit $\mathcal{E}_{in}$:

$$\mathcal{E}_{Nott} = \mathcal{E}_{em} - \mathcal{E}_{in}, \tag{2.111}$$

where one takes the value of the Fermi level energy W_F instead of $\mathcal{E}_{in}$. This is in fact true only in metals (and is not true in semiconductors, but the calculation of $\mathcal{E}_{in}$ is very nontrivial [7–9]). Hence we use approximate formula (2.117)

The thermal flow from the emitter point is equal to

$$Q_{em} = \mathcal{E}_{Nott}\, j_{em}, \tag{2.112}$$

where j_{em} is the emission current density. Formula (2.112) shows that if $\mathcal{E}_{Nott} > 0$, then the emitter is cooled, and if $\mathcal{E}_{Nott} < 0$, then the emitter is heated.

The average energy of emitting electrons is determined by the formula [39]

$$\mathcal{E}_{em} = \frac{e}{j_{TF}} \int_{-\infty}^{\infty} \widetilde{W} N_{\widetilde{W}}(\widetilde{W}, T, W_F) \int_{-\infty}^{\widetilde{W}} D(W)\,\mathrm{d}W\,\mathrm{d}\widetilde{W}, \tag{2.113}$$

where j_{TF} is the emission current density, see (2.102),

$$N_{\widetilde{W}}(\widetilde{W}, T, W_F) = -\frac{4\pi m_e}{h^3}\left(\exp\left(\frac{\widetilde{W} - W_F}{k_B T} \right) + 1 \right)^{-1}, \tag{2.114}$$

and $D(W)$ is the barrier transparency factor given by formula (2.88). The calculation of the "exact" value of $\mathcal{E}_{Nott}$ by the formulas under study results in the calculation of the current density j_{TF} and the double integral of composite functions containing elliptic integrals. These integrals can be calculated only by numerical methods, and such calculations are rather labor consuming. But there are approximate formulas for calculating the average energy $\mathcal{E}_{Nott}$, and the use of such formulas permits avoiding the calculation of integrals.

The following approximation for the average energy $\mathcal{E}_{Nott}$ was proposed in [34]:

$$\mathcal{E}_{FN} = -\pi k_B T \cot\left(\frac{\pi}{2}\frac{T}{T^*} \right), \tag{2.115}$$

where T^* is the *inversion temperature* which has the form [39]:

$$T^* = \frac{d_{FN}(E_F, \varsigma, \hat{y})}{2k_B} = \frac{\hbar e E_F}{4k_B\sqrt{2m_e}\,\varsigma t_{FN}(\hat{y}_s)}, \tag{2.116}$$

where ς is the work function, $d_{FN}(E_f, \varsigma, \hat{y}_s)$ is defined by (2.107), $t_{FN}(\hat{y}_s)$ is defined by (2.34), and $\hat{y}_s$ is defined by (2.106).

The inversion temperature is determined by the relation $\mathcal{E}_{Nott} = 0$. Therefore, in the interval where the temperature is lower than the inversion temperature, the emitter of electrons is heated ($\mathcal{E}_{Nott} < 0$), and in the temperature interval above T^*, the emitter cools down ($\mathcal{E}_{Nott} > 0$).

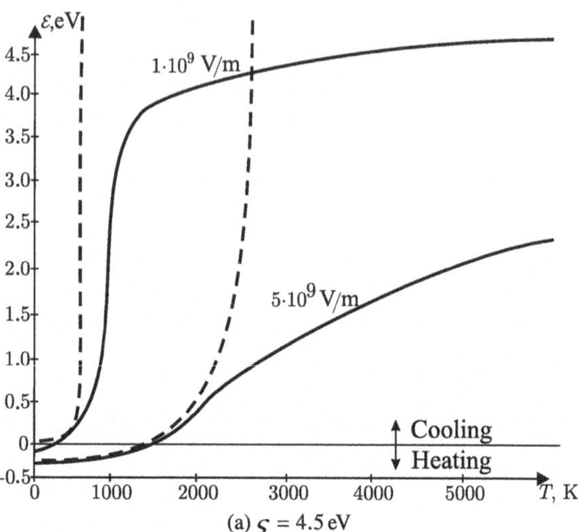

Fig. 2.17 Comparison of the exact value of the Nottingham effect (2.113) (solid line) with approximation (2.115) (dashed line) in [39]

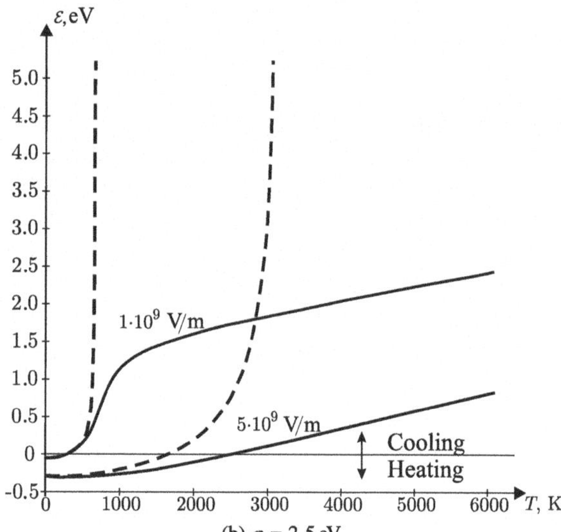

In Fig. 2.17a, the energy value $\mathcal{E}_{\text{Nott}}$ obtained by approximation (2.115) is compared with its exact value calculated by formula (2.113) for some fields. The approximation gives a quite satisfactory result for the temperatures $T \leqslant T^*$ but it may happen that this approximation is not true at the temperatures $T > T^*$ when cot is discontinuous at $T = 2kT^*$, $k = 1, 2, \ldots$. The approximation cannot also be used for small values of the work function and for very strong fields, see Fig. 2.17b.

By the Nottingham effect, the real values of heating and cooling are proportional to Q_{em}. Since $|\mathcal{E}_{\text{Nott}}|$ and j_{TF} are relatively small for $T < T^*$, but in the problem considered here they are large, we can conclude that we are in the domain $T > T^*$ and, respectively, the Nottingham effect in the problem under study is mainly the cooling effect. Obviously, approximation (2.115) is not true in the cooling domain $T > T^*$.

An improved approximation which holds for any temperatures of the emitter was proposed in [39]. This approximation has the form

$$\mathcal{E}_{\text{Fit}} = \begin{cases} -aa_T \dfrac{2}{\pi} T \cot\left(\dfrac{\pi}{2}\dfrac{T}{T^*}\right), & T \leqslant T^*, \\ aa_T(T - T^*)(1 - \vartheta) + (a_\infty T + b)\vartheta, & T > T^*, \end{cases} \tag{2.117}$$

where

$$a_{T^*} = \frac{1}{e}\frac{\mathrm{d}}{\mathrm{d}T}\mathcal{E}_{FN}\Big|_{T=T^*} = \frac{\pi^2 k_{\mathrm{B}}}{2e}, \tag{2.118}$$

$$a_\infty = \frac{1}{e}\frac{\mathrm{d}}{\mathrm{d}T}\mathcal{E}_{\text{em}}\Big|_{T\to\infty} = \frac{2k_{\mathrm{B}}}{e}, \tag{2.119}$$

$$\vartheta = \frac{p^\delta}{1 + p^\delta}; \quad p = \eta\left(\frac{T}{T^*} - 1\right). \tag{2.120}$$

An approximation for the inversion temperature was also proposed in [39]. It has the form

$$T^* = \omega_1 \frac{E_F^{\omega_2}}{\varsigma^{\omega_3}}\left\{1 + \frac{\omega_4}{\varsigma^{\omega_5}}\left[1 + \tanh\left(\omega_6 \frac{E_F^{\omega_7}}{\varsigma^{\omega_8}} - \omega_9\right)\right]\right\}, \tag{2.121}$$

where the parameters $\omega_1, \ldots, \omega_9$ taken from [39] are given in Sect. 2.4.8. The comparison of (2.121) with exact formula (2.116) for an example of silicon emitter is given in Sect. 2.4.9.

The Nottingham effect for the field emission precisely from semiconductors was considered in [8, 9], where the theory of replacement of the emitted electrons by electrons from the circuit in semiconductor emitters was developed under the assumption that the electron coming from the circuit can occupy any empty place in the conduction band or in the valence band. These works also present formulas for the average energy of electrons leaving the cathode $\mathcal{E}_{\text{em}}$ and for the average energy of electrons approaching the emitter surface from the electric circuit $\mathcal{E}_{\text{in}}$:

$$\mathcal{E}_{\text{em}} = \frac{\int\limits_{-\infty}^{\infty} W P_e(W)\,\mathrm{d}W}{\int\limits_{-\infty}^{\infty} P_e(W)\,\mathrm{d}W}; \quad \mathcal{E}_{\text{in}} = \frac{\int\limits_{-\infty}^{\infty} W P_r(W)\,\mathrm{d}W}{\int\limits_{-\infty}^{\infty} P_r(W)\,\mathrm{d}W}, \tag{2.122}$$

where the integrands $P_e(W)$ and $P_r(W)$ are distinct for semiconductors of different types; namely, for the p-type semiconductors, they have the form

$$P_e^p(W) = A_{\mathrm{Not}} \int_W^{W_v} (1 - f_{\mathrm{FD}}(W))D(W_x)\, \mathrm{d}W_x; \tag{2.123}$$

$$P_r^p(W) = A_{\mathrm{Not}} \int_W^{W_v} (1 - f_{\mathrm{FD}}(W))\left[f_{\mathrm{FD}}(W) + (1 - f_{\mathrm{FD}}(W))D(W_x)\right] \mathrm{d}W_x, \tag{2.124}$$

and for the n-type semiconductors, the form

$$P_e^n(W) = A_{\mathrm{Not}} \int_{W_c}^{W} f_{\mathrm{FD}}(W)D(W_x)\, \mathrm{d}W_x; \tag{2.125}$$

$$P_r^n(W) = A_{\mathrm{Not}} \int_{W_c}^{W} f_{\mathrm{FD}}(W)\left[1 - f_{\mathrm{FD}}(W) + f_{\mathrm{FD}}(W)D(W_x)\right] \mathrm{d}W_x, \tag{2.126}$$

where W_c is the lower bound of the conduction band, W_v is the upper bound of the valence band (see Fig. 2.2a in Sect. 2.1), $D(W)$ is the barrier transparency factor (see (2.88)), $f_{\mathrm{FD}}(W)$ is the Fermi–Dirac distribution function (see (2.2)), and

$$A_{\mathrm{Not}} = \frac{4\pi m_e e}{h^3}.$$

But these formulas (2.122)–(2.126) hold for the field emission and it is unknown whether they hold for the thermo-field emission. By this reason and because it is rather difficult to calculate by these formulas, we do not use them in numerical modeling (see Chap. 4) but use approximate formula (2.117).

2.4.8 Optimal Values of Approximation Parameters

In this Section, we present optimal values of the parameters for the functions (2.109), (2.108), (2.110), (2.121), (2.117) which are obtained in [39]. In what follows, we also assume that E_{F} is measured in B/m, and ς, in eV. The maximal error $r_{\max}$ is the maximal deviation of the approximation from the exact formula, and the mean error $\bar{r}$ is, respectively, the mean deviation of the approximation from the exact formula on the entire domain of definition.

1. In formula (2.108), the optimal parameter is

$$\beta_y = 1.65516 \tag{2.127}$$

with $\bar{r} = 0.67\%$ and $r_{\max} = 1.28\%$.

2. In formula (2.109), the first set of optimal parameters has the form

$$\alpha_1 = 0.109392; \quad \beta_1 = 1.21980, \tag{2.128}$$

and using it, we obtain $\bar{r} = 0.19\%$ and $r_{\max} = 0.45\%$.

3. In formula (2.109), the second set of optimal parameters has the form

$$\alpha_2 = 0.231723; \quad \beta_2 = 1.29655, \tag{2.129}$$

and using it, we obtain $\bar{r} = 0.27\%$ and $r_{\max} = 0.64\%$.

4. The approximation for the formula of the thermo-field emission current has the form (2.110), and the optimal values of the parameters q_i are

$$
\begin{aligned}
&q_1 = 7.32527 \times 10^{-6}; &&q_2 = 1.90593; \\
&q_3 = 0.810323; &&q_4 = 2.36641 \times 10^{-22}; \\
&q_5 = 2.43459; &&q_6 = 5.22916; \\
&q_7 = 6.71665 \times 10^{9}; &&q_8 = 1.49781; \\
&q_9 = 0.998795; &&q_{10} = 197.477; \\
&q_{11} = 0.205375; &&q_{12} = 0.145045.
\end{aligned}
$$

The approximation was performed for the values

$$E_{\mathrm{F}} \in [3.16 \times 10^{7}, 10^{10}]\,\text{V/m and } \varsigma \in [2, 5]\,\text{eV}.$$

In this interval, the mean error is $\bar{r} = 0.85\%$ and the maximal error is $r_{\max} = 36\%$.

5. As was already shown, the approximation for the inversion temperature has the form (2.121), where the optimal values of the parameters ω_i are

$$
\begin{aligned}
&\omega_1 = 7.1130 \times 10^{-7}; &&\omega_2 = 0.98604; \\
&\omega_3 = 0.47483; &&\omega_4 = 1.0296; \\
&\omega_5 = 0.91905; &&\omega_6 = 4.8022; \\
&\omega_7 = 8.8832 \times 10^{-2}; &&\omega_8 = 0.15358; \\
&\omega_9 = 30.371.
\end{aligned}
$$

The approximation was performed for the values

$$E_{\mathrm{F}} \in [2 \times 10^{8}, 10^{10}]\,\text{V/m and } \varsigma \in [2, 5]\,\text{eV}.$$

In this interval, we have the mean error $\bar{r} = 1.1\%$ and the maximal error $r_{\max} = 4.56\%$. Equation (2.121) also holds for fields of strength less than 2×10^{8} V/m.

6. In formula (2.117), the optimal values of the parameters are

$$a = 1 - 1.03104 \times 10^{-2} E_{\mathrm{F}}^{0.193326} / \varsigma^{0.821433};$$

$$b = \varsigma - 1.99435 \times 10^{-5} E_{\mathrm{F}}^{0.533739};$$

$$\eta = 0.687365/\varsigma^{0.0525966};$$

$$\delta = 3.48481.$$

2.4.9 Dependence of the Inversion Temperature on the External Electric Field Strength

Now we illustrate the calculation of the inversion temperature T^* in more detail with an example of silicon cathode. Here we can use the following two approaches: "exact" Fowler–Nordheim formula (2.116) and approximate formula (2.121). It should be noted that, in formula (2.116), the work function is given in Joules, and in formula (2.121), in electronvolts.

In the calculations by exact formula (2.121), we used the values of physical constants given in Table 2.1. The work function in silicon is $\varsigma = 4.8\,\mathrm{eV}$ [22]. As an approximation for $t_{\mathrm{FN}}(y)$ we took formula (2.109) with parameters (2.128). When calculating by approximate formula (2.121), we used the parameters $\omega_1, \ldots, \omega_9$ given in this section.

In Fig. 2.18a, one can see the dependence of the inversion temperature on the value of the external field calculated by Fowler–Nordheim formula (2.116) and by approximate formula (2.121). One can see that, for the fields 6×10^8–5×10^9 V/m, the approximate formula gives a rather good approximation, but for stronger fields, the approximate formula gives a large discrepancy.

In Fig. 2.18b, one can see the dependence of the inversion temperature in the range of fields 10^6–5×10^8 V/m that are typical of the silicon emitter regime considered here. The values of the inversion temperature calculated by exact formula (2.116) weakly differ from the values calculated by formula (2.121), and for small values of strength, less than by 1 K. For example, for $E_{\mathrm{F}} = 10^8$ V/m,

$$T^*_{\mathrm{exact}} = 25.7709\,\mathrm{K}, \quad T^*_{\mathrm{approx}} = 26.1115\,\mathrm{K},$$

Table 2.1 Physical constants [22]

m_e	electron mass	9.109×10^{-31} kg
e	electron charge	1.602×10^{-19} C
ς	work function (for silicon)	$4.8\,\mathrm{eV}$
k_{B}	Boltzmann constant	1.381×10^{-23} J/K
$\hbar$	Planck constant	1.055×10^{-34} J·s
ε_0	permittivity	8.854×10^{-12} F/m

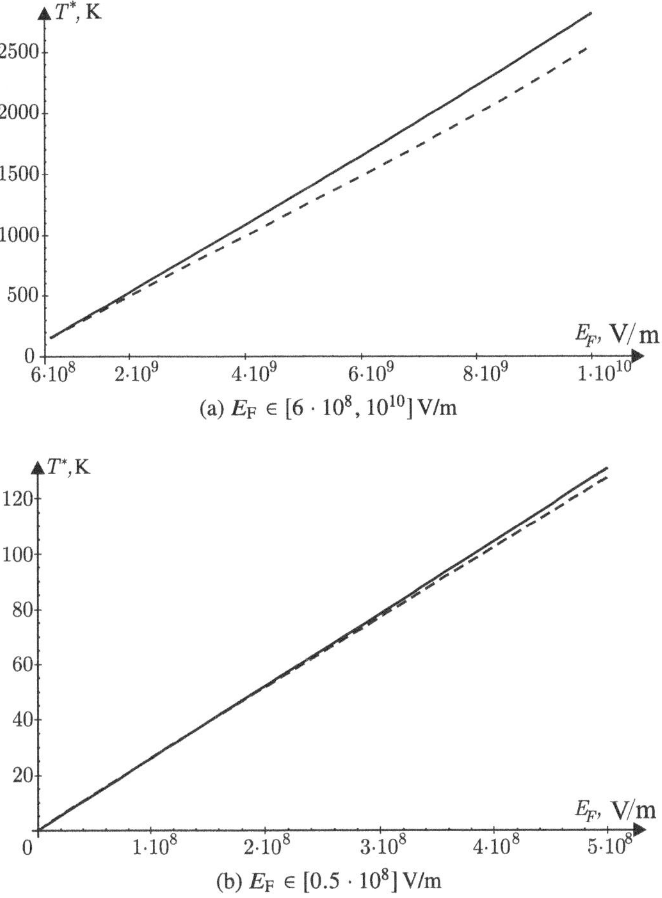

Fig. 2.18 Comparison of the dependence of inversion temperature on the external electric field for $\varsigma = 4.8\,\text{eV}$ and T^*_{exact} calculated by Fowler–Nordheim formula (2.116) (solid line) with T^*_{approx} calculated by approximate formula (2.121) (dashed line)

and for $E_F = 2 \times 10^8\,\text{V/m}$,

$$T^*_{\text{exact}} = 51.4439\,\text{K}, \quad T^*_{\text{approx}} = 51.7281\,\text{K}.$$

The approximate formula gives an overestimated value but such an error does not significantly influence the results of numerical experiment.

The main conclusion in this section is the following one. In the range of electric field strength under study, the inversion temperature T^* of the silicon emitter is rather low ($T^* < 200\,\text{K}$). This means that, considering the regime of cathode operation near the "melting–crystallization" point (the silicon melting temperature is $\approx 1700\,\text{K}$), we necessarily are in the temperature range $T > T^*$. Thus, the Nottingham effect in our problem always means that the emission surface of the cathode is cooled.

In semiconductors, in contrast to metals, the conductivity strongly increases with the temperature. In the transition to liquid state in semiconductors, there arise additional electrons (due to the breakage of covalent bonds between atoms). Therefore, in the region of local melting, the conductivity increases jumpwise, and this can decrease the Joule heat. But the exact value of variation in the conductivity has not been determined. On the other hand, for general reasons it is natural to assume that a part of free electrons from the melting zone can increase the emission current density, which results in an increase in the Nottingham effect, i.e., in the energy transferred away by emitting electrons. In the above model, we use the formula describing the Nottingham effect for the emission of electrons from metals, which does not take into account the above-mentioned increase in the flow of emitting electrons. Therefore, we introduce an artificial "form-factor", i.e., a correction coefficient, in this formula, which allows us to take into account both of the (hypothetical) phenomena described above, because both of them lead to the cooling of the cathode material.

References

1. Abramowitz, M., Stegun, I.A. (eds.): Handbook of Mathematical Functions with Formulas, Graphs, and Mathematical Tables. Dover Books on Mathematics. Dover Publications, NY (1965)
2. Ashcroft, N.W., Mermin, N.D.: Solid State Physics. Cengage Learning (1976)
3. Bonch-Bruevich, V.L., Kalashnikov, S.G.: Semiconductor Physics. Nauka, Moscow (1990). (in Russian)
4. Charbonnier, F.M., Strayer, R.W., Swanson, L.W., Martin, E.E.: Nottingham effect in field and t-f emission: heating and cooling domains, and inversion temperature. Phys. Rev. Lett. **13**(13), 397–401 (1964)
5. Christov, S.G.: General theory of electron emission from metals. Phys. Status Ssolidi (B) **17**(1), 11–26 (1966)
6. Chung, M.S., Cutler, P.H., Miskovscky, N.M., Sullivan, T.E.: Energy exchange processes in electron emission at high fields and temperature. J. Vacuum Sci. Technol. B **12**(2), 727–736 (1994)
7. Chung, M.S., Hyun, S.S.: Derivation of the average energy of the field electrons emitted from semiconductors. J. Korean Phys. Soc. **38**(6), 758–761 (2001)
8. Chung, M.S., Jang, Y.J., Mayer, A., Cutler, P.H., Miskovscky, N.M., Weis, B.L.: Energy exchange in field emission from semiconductors. J. Vacuum Sci. Technol. B **26**, 800–805 (2008)
9. Chung, M.S., Jang, Y.J., Mayer, A., Cutler, P.H., Miskovscky, N.M., Weis, B.L.: Theoretical analysis of the energy exchange and cooling in field emission from the conduction band of the n-type semiconductor. J. Vacuum Sci. Technol. B **27**, 692–697 (2009)
10. Ding, M.: Field emission from silicon: Ph.D. thesis. Massachusetts Institute of Technology (2001)
11. Egorov, N., Sheshin, E.: Field Emission Electronics. Springer (2017)
12. Elinson, M.I., Vasil'ev, G.F.: Field Emission. Fizmatgiz, Moscow (1958). (in Russian)
13. Fleming, G.M., E., H.J.: The energy losses attending field current and thermoionic emission of electrons from metals. Phys. Rev. **58**, 887–894 (1940)

14. Flügge, S. (ed.): Electron-Emission and Gas Discharges I, Encyclopedia of Physics, vol. XXI. Springer, Berlin (1956)

15. Forbes, R.G.: Simple good appeoximations for the special elliptic functions in standart fowler-nordheim tunneling theory for a Schottky-Nordheim barrier. Appl. Phys. Lett. **89** (2006)

16. Forbes, R.G.: On the need for a tunneling pre-factor in Fowler-Nordheim tunneling theory. J. Appl. Phys. **103** (2008)

17. Forbes, R.G., Deane, J.H.B.: Reformulation of the standart theory of Fowler-Nordheim tunneling and cold field electron emission. Proc. Royal Soc. A **463**, 2907–2927 (2007)

18. Fowler, R.H., Nordheim, L.: Electron emission in intense electric fields. Proc. Royal Soc. Lond. Ser. A **119**(781), 173–181 (1928)

19. Fröman, H., Fröman, P.O.: JWKB Approximation: Contributions to the Theory. North-Holland Pub, Amsterdam (1965)

20. Furcey, G.: Field Emission in Vacuum Microelectronics. Springer (2005)

21. Glazov, V.M., Chizhevskaia, S.N., Glagoleva, N.N.: Liquid Semiconductors. Springer (1969)

22. Grigoriev, I.S., Meilikhov, E.Z., Radzig, A.A.: Handbook of Physical Quantities. CRC Press (1997)

23. Grundmann, M.: The Physics of Semiconductors: An Introduction Including Nanophysics and Applications. Springer (2016)

24. Hantzsche, E.: Theory of cathode spot phenomena. Physica B+C **104**, 3–16 (1981)

25. Hantzsche, E.: The thermo-field emission of electrons in arc discharges. Beiträge aus der Plasmaphysik **22**(4), 325–346 (1982)

26. Hantzsche, E.: The state of the theory of vacuum arc cathodes. Beiträge aus der Plasmaphysik **23**(1), 77–94 (1983)

27. Hofmann, P.: Solid State Physics : An Introduction. Wiley (2015)

28. Kasap, S., Capper, P. (eds.): Springer Handbook of Electronic and Photonic Materials. Springer, US (2007)

29. Kemble, E.C.: The Fundamental Principles of Quantum Mechanics. McGraw-Hill, NY (1937)

30. Kittel, C.: Introduction to Solid State Physics. Wiley (2005)

31. Lafferty, J.M. (ed.): Vacuum Arcs: Theory and Application. Wiley, NY (1980)

32. Landau, L.D., Lifshitz, E.M.: Quantum Mechanics-Nonrelativistic Theory (Course of Theoretical Physics). Pergamon Press (1981)

33. Lee, T.H.: T-f theory of electron emission in high-current arcs. J. Appl. Phys. **30**(2), 166–171 (1959)

34. Levine, P.H.: Thermoelectric phenomena associated with electron-field emission. J. Appl. Phys. **33**(2), 582–587 (1962)

35. Miller, S.C., Good, R.H.: A WKB-type approximation to the Schrödinger equation. Phys. Rev. **91**(1), 174–179 (1953)

36. Modinos, A.: Field, Thermionic, and Secondary Electron Emission Spectroscopy. Springer, US (1984)

37. Murphy, E.L., Good, R.H.: Thermionic emission, field emission, and the transition region. Phys. Rev. **102**(6), 1464–1473 (1956)

38. Nottingham, W.B.: Remarks on energy losses attending thermionic emission of electrons from metals. Phys. Rev. (1941)

39. Paulini, J., Klein, T., Simon, G.: Thermo-field emission and the Nottingham effect. J. Phys. D: Appl. Phys. **26**(8), 1310–1315 (1993)

40. Richardson, O.: Thermionic phenomena and the laws which govern them. In: Nobel Lecture, pp. 224–236. Stockholm (1929)

41. Shalimova, K.V.: Physics of Semiconductors. Energoatomizdat, Moscow (1985). (in Russian)

42. Stilbans, L.S.: Physics Semiconductors. Soviet radio, Moscow (1967). (in Russian)

43. Stratton, R.: Theory of field emission from semiconductors. Phys. Rev. **125**(1), 67–82 (1962)

44. Vainshtein, I.A., Zatsepin, A.F., Kortov, V.S.: Applicability of the empirical varshni relation for the temperature dependence of the width of the band gap. Phys. Solid State **41**(6), 905–908 (1999)

45. Vallée, O., Soares, M.: Airy Functions and Applications to Physics. Imperial College Press, London (2004)
46. Varshni, Y.P.: Temperature dependence of the energy gap in semiconductors. Physica **34**(1), 149–154 (1967)

Chapter 3
Mathematical Model

Abstract This chapter is a "mathematical" one. Here we collect the mathematical background related to the mathematical model of phase transition based on the phase field system introduced by G. Caginalp. Sections 3.1 and 3.2 of the chapter contain some preliminaries and considerations about mathematical models from the physical viewpoint. In Sect. 3.3, we give the results of asymptotic analysis applied to the phase field system. In Sect. 3.4, we discuss a new definition of the generalized solution to the phase field system which is stable under passing to the limiting Stefan–Gibbs–Thomson problem. Finally, in Sect. 3.5, we discuss an approach which is a combination of mathematical (asymptotic) investigation and numerical analysis.

3.1 Phase Field System and its Use in Heat Transfer Modeling

In this section, we give general information about the problems with free boundary (namely, the phase transition) and their regularization. One of the most important applications of such problems is the simulation of melting—solidification processes. We here discuss the temperature and other parameters without mentioning their specific values and focus our attention on the mathematical background of the problem.

We consider a matter which can be in two states (two phases): solid and liquid. Assume that this matter occupies a domain $\Omega \subset \mathbb{R}^3$. We denote the dimensionless temperature by θ and choose the reference point for the temperature so that $\theta = 0$ is the melting temperature of the matter.

In the classical Stefan problem (this is historically the first mathematical model describing the matter evolution in the presence of two phases), it is assumed that the interface between the phases Γ is determined as

$$\Gamma(t) = \left\{ x \in \Omega,\ \theta(x, t) = 0 \right\}. \tag{3.1}$$

© Springer Nature Singapore Pte Ltd. 2020

V. Danilov et al., *Mathematical Modeling of Emission in Small-Size Cathode*, Heat and Mass Transfer, https://doi.org/10.1007/978-981-15-0195-1_3

Here and below, x is the dimensionless coordinate, and t is the dimensionless time. The domain occupied by the liquid phase Ω_{liq} is determined by the relation

$$\Omega_{\text{liq}} = \{x \in \Omega, \ \theta(x, t) > 0\}, \tag{3.2}$$

and the domain occupied by the solid phase, by the relation

$$\Omega_{\text{sol}} = \{x \in \Omega, \ \theta(x, t) < 0\}. \tag{3.3}$$

The (dimensionless) temperature θ must satisfy the heat conduction equation

$$\theta_t = k\Delta\theta, \tag{3.4}$$

where k is the dimensionless thermal conductivity. Moreover, the Stefan condition must be satisfied on the boundary:

$$lv_{\mathbf{n}} = k\left[\frac{\partial\theta}{\partial\mathbf{n}}\right]\bigg|_{\Gamma(t)}. \tag{3.5}$$

Here

$$\left[\frac{\partial\theta}{\partial\mathbf{n}}\right]\bigg|_{\Gamma(t)} = \langle\nabla\theta_{\text{sol}} - \nabla\theta_{\text{liq}}, \mathbf{n}\rangle, \quad x \in \Gamma(t),$$

where l is the dimensionless melting heat, $v_{\mathbf{n}}$ is the normal velocity of points of the free boundary $\Gamma(t)$, $\mathbf{n}$ is the normal to $\Gamma(t)$ (in the direction from the solid phase to the liquid phase).

The conditions typical of the heat equation are posed on the outer boundary and for $t = 0$: for example, the Dirichlet condition

$$\theta(x, t) = \theta_\partial(x, t), \quad x \in \partial\Omega, \ t \in \mathbb{R}^1, \tag{3.6}$$

or the boundary condition of the third kind and the initial condition

$$\theta(x, 0) = \theta_0(x), \quad x \in \Omega. \tag{3.7}$$

Mathematically, this problem is to determine the function $\theta(x, t)$ and the free boundary $\Gamma(t)$.

Thus, the solution of Stefan problem (3.1), (3.4)–(3.7) is a pair, i.e., the function $\theta(x, t)$ and the surface $\Gamma(t)$. The fact that it contains an unknown surface makes the Stefan problem strongly nonlinear despite the fact that the expressions containing the unknown temperature $\theta(x, t)$ are formally linear (for a given surface $\Gamma(t)$).

But it turns out that this sufficiently simply formulated mathematical model is rather far from the physical reality, especially, if the system under study cannot be treated as one-dimensional. The Stefan model does not take into account important physical information such as the surface tension on the interface between the phases.

As Gibbs [30] noted, a direct consequence of the fact that the surface tension was taken into account as an additional stability factor is the following modification of the Stefan problem condition (3.1) $\left(\theta\big|_{\Gamma(t)} = 0\right)$ on the free boundary so that it becomes [2, 57]:

$$\Delta s\, \theta(x, t) = -\beta \mathcal{K}(x, t), \quad x \in \Gamma(t). \tag{3.8}$$

Here Δs is the difference of entropies of the liquid and solid phases, $\mathcal{K}(x, t)$ is the sum of principal curvatures of the surface (mean curvature) at the point $x \in \Gamma(t)$, and the coefficient β depending on the (cathode) material has the form

$$\beta = \frac{\sigma \theta_0}{\rho \tilde{l}},$$

where σ is the surface tension, θ_0 is the (dimensionless) melting temperature, ρ is the density, and $\tilde{l}$ is the latent heat of melting, see Table 4.1 in Sect. 4.1 for details.

Condition (3.8) readily implies that if the interface between phases is plane, then we return to condition (3.1). In addition to the effect arising due to the surface tension taken into account, the temperature at the points of $\Gamma(t)$ must further decrease (or increase); this is called the overheating (or overcooling) effect [15]. To take this effect into account, one can use different expressions, but the most frequently used method is to introduce a linear dependence on the rate of increase (decrease) [39] into (3.8):

$$\Delta s\, \theta(x, t) = -\beta \mathcal{K}(x, t) - \alpha v_{\mathbf{n}}(x, t), \quad x \in \Gamma(t). \tag{3.9}$$

Here the coefficient α depending on the cathode material has the form [39]:

$$\alpha = \frac{c}{\mu \tilde{l}},$$

where c is the specific heat capacity and μ is the kinetic coefficient of growth, see Table 4.1 in Sect. 4.1. for details.

We note that the interface between the phases in the Stefan problem is unstable [26]. The change of (3.1) by (3.8) must, obviously, improve the stability, because condition (3.8) restricts the possible curvature of the surface (and the instability starts precisely from local perturbations of the mean curvature). Thus, one can say that, for large values of the surface tension σ, the local instabilities on $\Gamma(t)$ do not develop but decay.

The coefficient α in condition (3.9) (and in general, the term containing the rate of growth) does not act in this way; this is only a stabilizing factor. The linear analysis [9] shows that unstable modes in the Stefan problem remain unstable if condition (3.1) is replaced by condition (3.9) with $\beta = 0$, but the amplitude of these modes increases slower.

Thus, conditions (3.1), (3.8), and (3.9) result in different qualitative properties of the interface evolution type. For a given material, one can hope to obtain an adequate result for an appropriate definition of the parameters α and β in condition (3.9). It is clear that condition (3.9) (and condition (3.8)) make the model more nonlinear than the Stefan model. The model with general condition (3.9) is called the *Stefan–Gibbs–Thomson problem*.

Now we reformulate the Stefan–Gibbs–Thomson problem in the language of generalized functions. Note that heat equation (3.4) is understood in the sense of classical analysis, i.e. the partial derivatives of the temperature are considered on one or other side from the interface but not at the points of the surface itself. It is easy to see that Eq. (3.4) and condition (3.5) can be united in a single equation, where the derivatives are already understood in the weak sense (as in the theory of generalized functions), namely,

$$\theta_t + \frac{l}{2}\bar{\varphi}_t = k\Delta\theta, \quad \bar{\varphi} = \begin{cases} -1, & x \in \Omega_{\text{sol}}, \\ +1, & x \in \Omega_{\text{liq}}. \end{cases} \tag{3.10}$$

Then one can say that $\Gamma(t)$ is the surface on which the function $\bar{\varphi}$ defined in the domain $\Omega \times \mathbb{R}^1$ is discontinuous.

To clarify the last relation, we consider the one-dimensional case. In this case, $\Gamma(t) = \{x = \psi(t)\}$,

$$\theta(x,t) = \theta_{\text{sol}}(x,t) + H(x - \psi(t))\big(\theta_{\text{liq}}(x,t) - \theta_{\text{sol}}(x,t)\big), \quad \varphi(x,t) = \text{sign}\,(x - \psi(t)),$$

where $H(z)$ is the Heaviside function and $\text{sign}\,(z) = H(z) - H(-z)$. For a moment, we assume that $\theta_{\text{sol}}, \ \theta_{\text{liq}} \in C^2, \ \psi(t) \in C^1$, and $(\theta_{\text{sol}} - \theta_{\text{liq}})\big|_{x=\psi} = 0$. Then, substituting the above expressions into (3.10) and calculating the coefficient at $\delta(x - \psi)$, we obtain

$$\delta(x - \psi)\left(-l\psi_t - k\left(\frac{\partial\theta_{\text{sol}}}{\partial x} - \frac{\partial\theta_{\text{liq}}}{\partial x} \right)\bigg|_{x=\psi} \right) + \text{ less singular summunds} = 0.$$

Note that $\psi_t = v$ is the velocity of the point $x = \psi(t)$. From the last relation we have (3.5) on the one-dimensional case. The same is true for arbitrary spatial dimensions.

It is clear that such a statement of the problem is related to condition (3.1) and the change of condition (3.1) by condition (3.9) requires certain changes in the definition of the function $\bar{\varphi}$, because the temperature on $\Gamma(t)$ need not now be equal to zero. It is necessary to introduce this function in condition (3.9). Here one can proceed in the following two ways.

The first method (conditionally, physical) is to construct a model of the phase transition process which explicitly contains the function $\bar{\varphi}$ or its analog and makes it physically meaningful.

The second method is to consider problem (3.4)–(3.7), (3.10) as the limiting problem by letting the small parameter ε to tend to zero in the regularized problem, i.e., by regularizing problem (3.4)–(3.7), (3.10). An analogy known in this case is the correspondence between the limit Hopf equation

$$u_t + \left(u^2\right)_x = 0 \tag{3.11}$$

and its regularization

$$u_t^\varepsilon + \left(\left(u^\varepsilon\right)^2\right)_x = \varepsilon u_{xx}^\varepsilon, \tag{3.12}$$

which are the essence of the small viscosity method in the theory of conservation laws.

First, we briefly consider the first method. In the physical literature, there is much information about the behavior of nonequilibrium thermodynamic systems: the Ginzburg–Laudau model, the Landau–Khalatnikov model, and many others. It is assumed in these models that the state of the system is determined not only by physical parameters, such as temperature, density, and so on, but also by an additional dimensionless "order parameter" $\varphi = \varphi(x, t)$ (or by several of such parameters, see, e.g., model A in [33]). In this context, the following system of equations was considered:

$$\begin{cases} \theta_t + \dfrac{l}{2}\varphi_t = k\Delta\theta; \\[2mm] \alpha\xi^2\varphi_t = \xi^2\Delta\varphi + \dfrac{1}{a}g(\varphi) + 2\theta, \end{cases} \tag{3.13}$$

where $g(\varphi)$ is the derivative of the potential $W(\varphi)$ that is symmetric with respect to zero and has two maxima at the points $\varphi = \pm 1$. In the simplest case,

$$g(\varphi) = \frac{\varphi - \varphi^3}{2}.$$

The parameters l, k, α, ξ, and a are dimensionless combinations of physical parameters of problem [5, 6]. We discuss their definitions in more detail below. One can see that the first equation in (3.13) coincides in form with Eq. (3.10), but the solution of the equation in system (3.13) has no jump like the function $\bar\varphi$ in (3.10). If we formally put

$$a = \Delta s\big|_{\Gamma(t)}, \quad a\xi^2 = \sigma\varepsilon^2 \tag{3.14}$$

and assume that ε is a formal parameter, then in the limit as $\varepsilon \to 0$, the second equation in (3.13) becomes (3.9).

The main idea of the second approach is to replace the original Stefan–Gibbs–Thomson problem, which is not the "standard" boundary-value problem in the theory of differential equations because of the presence of an unknown free boundary, by a new problem, i.e., problem (3.13) in our case, which already belongs to the standard class of boundary-value problems for (nonlinear) differential equations. This new "standard" problem depends on a small parameter ε, and passing to the limit with respect to this parameter in the "standard" solution gives precisely the solution of the original "nonstandard" problem. This is the key point of our algorithm [18, 20, 22, 23].

We note that the relationship between Eqs. (3.11) and (3.12) is formally clear: by setting $\varepsilon = 0$ in (3.12), we obtain (3.11). But it only seems that this is so simple. In fact, when passing from (3.12) to (3.11), we pass to the limit as $\varepsilon \to 0$ and the main difficulty is to verify the relation

$$\lim_{\varepsilon \to 0} \left(u^\varepsilon\right)^2 = \left(\lim_{\varepsilon \to 0} u^\varepsilon\right)^2,$$

which is one of the main problems in the theory of hyperbolic conservation laws. This theory has already been known for more than a century. The theory of passing to the limit in problems with a free boundary has completely been constructed in the last thirty years and is more complicated. At the first glance, there is at all no relationship between (3.13) and (3.10) or between (3.4), (3.5) and (3.8). But it exists, and this relationship is revealed in Sects. 3.3 and 3.4.

Here there is one significant detail, namely, since the "standard" problem is rather complicated, we cannot determine its exact solution and have to consider asymptotic approximations which can be unstable. The problem of this possible instability has not yet been solved theoretically. But in some specific systems, for example, in the case of problem (3.13), instead of theoretical study, one can use the results of numerical solution and verify the stability by varying the parameters of numerical experiments. On the one hand, this of course does not rigorously prove that the computer data thus obtained for small values of the regularization parameter actually approximate the exact solution of the Stefan–Gibbs–Thomson problem in the case of cathode melting. On the other hand, since this problem is very complicated, one should not at all expect to obtain a rigorous mathematical justification, and on the contrary, one needs to trust the results of numerous computer experiments for verifying the results, i.e., to restrict the level of rigor by developing an advanced "mathematical technology" rather than a complete mathematical theory. We note that there exists a rigorous justification of passing to the limit from phase field system (3.13) to the Stefan–Gibbs–Thomson problem without studying the problem of stability, but this justification is mathematically nontrivial and has been done in a simpler situation.

3.2 Phase Field System as Regularization of Limiting Problems with Free Boundary

In this section, we discuss sufficiently general problems of the possibility of using the phase field system to regularize problems of different types with free boundary. Here we follow the theory presented in [8]. As was already noted (see (3.13)), one can use the Ginzburg–Landau theory of phase transitions [33] to write the phase field system in the form [5, 6]

$$\theta_t = k\Delta\theta - \frac{l}{2}\varphi_t; \tag{3.15}$$

$$\alpha\xi^2\varphi_t = \xi^2\Delta\varphi + \frac{1}{a}g(\varphi) + 2\theta. \tag{3.16}$$

Here θ is the dimensionless temperature ($\theta = 0$ is the melting temperature), φ is the order function (phase field), g is the derivative of the potential that is symmetric with respect to zero and has two maxima at the points ± 1, i.e.,

$$g(\varphi) = \frac{\varphi - \varphi^3}{2}. \tag{3.17}$$

The parameters k, l, α, ξ, and a are some dimensionless constants. We note that, on the free boundary, one need not pose conditions (of Stefan- and Gibbs–Thomson-type which ensure the free boundary motion and describe the influence of the curvature and the interface velocity on the temperature at the points of free boundary), because these conditions can be obtained from the system of phase field equations (3.15), (3.16) directly as consequences of microscopic physical effects contained in these equations [3, 5–7, 11, 12].

We also note that, under sufficiently general conditions, there exists a unique smooth global solution of initial boundary-value problem for the system of phase field equations (3.15), (3.16) in the case of arbitrary dimension [6]. At the same time, there does not exist any theory of the existence of a solution for limiting problems with a rapidly varying localized perturbation on the free boundary (problems with distinct interface between phases, i.e., Stefan- and Hele-Shaw-type problems) [4, 32]. Here we show that the most known problems with distinct interface arise as the limit cases of the system of phase field equations (3.15), (3.16) in the asymptotic analysis in the case as $\xi, a \to 0$, and in some cases, also as $\alpha \to 0$.

Namely, a variation in the values of the parameters of the system of phase field equations (3.15), (3.16) (especially, a) significantly affects the limiting properties of these equations. In particular, one can obtain different limiting problems with different conditions on the free boundary stipulated by the different physical interpretation of values of the system parameters.

The main result of the asymptotic analysis is the fact that any Stefan- and Hele-Shaw-type model with any set of physical parameters and in the case of any dimension can be approximated with an arbitrary accuracy by an appropriate choice of phase field equations (3.15), (3.16) and conversely; see Figs. 3.1, 3.2 and [8].

The system of phase field equations is of great mathematical interest because one can use the same system of equations to study (for example, numerically) different physical processes such as the flows in Hele-Shaw-type problems or the solidification (melting) effects in Stefan-type problems. Varying the values of three parameters contained in system (3.15), (3.16), we can get different limiting problems.

We note that, in the asymptotic analysis, one can also obtain limiting problems consisting of a single equation. For example, by setting $l = 0$ and choosing zero

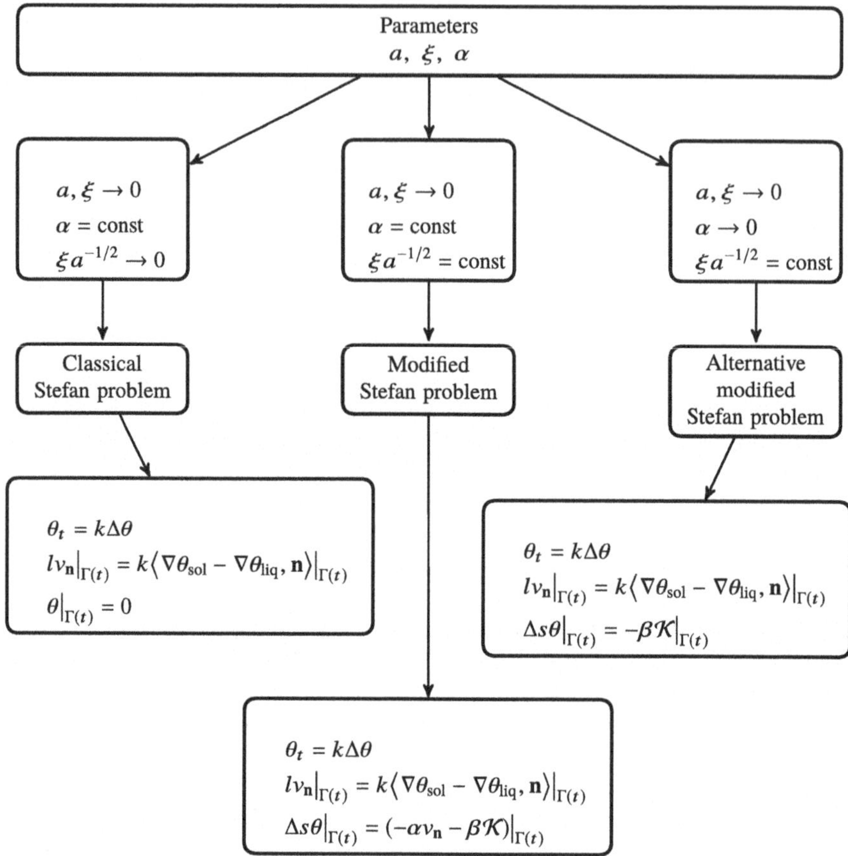

Fig. 3.1 Stefan-type models as the limit of the system of phase field equations (3.15), (3.16). Three main relations between the parameters of the system of phase field equations which lead to different limit conditions on the free boundary

initial conditions for the temperature (thus eliminating its influence), one can obtain the Allen–Cahn equation for the order function in the limit:

$$\alpha \xi^2 \varphi_t = \xi^2 \Delta \varphi + \frac{1}{a} g(\varphi). \tag{3.18}$$

Figures 3.1 and 3.2 show the limit transitions from the system of phase field equations to different limiting problems with free boundary for a certain choice of values of the system parameters.

We consider the basic (heuristic) ideas. Let r be the normal coordinate on the interface between phases $\Gamma(t)$ (i.e., let r be the distance to the interface if we are in the liquid domain, and let it be a negative distance if we are in the solid domain). We assume that in the initial boundary-value problem for the system of phase field

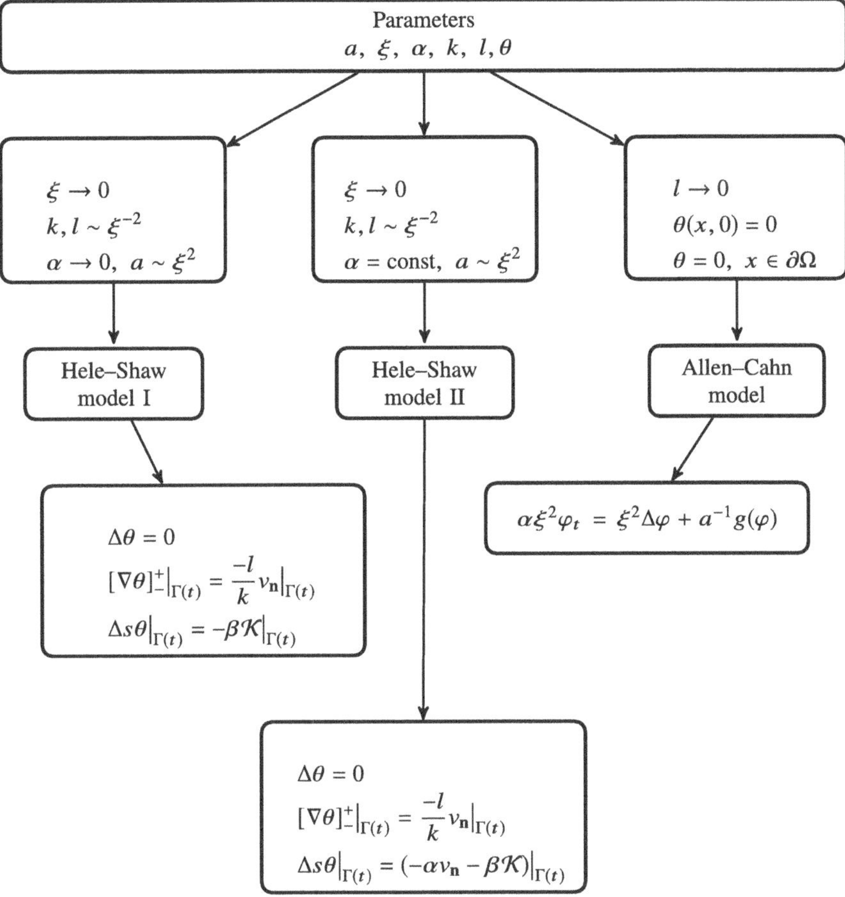

Fig. 3.2 Hele-Shaw- and Allen–Cahn-type models as the limit systems of phase field equations (3.15), (3.16). Versions of the relation between the parameters of the system of phase field equations which lead to principal distinctions in the properties of limiting problems

equations (3.15), (3.16), the function φ varies along the interface between the phases much faster than θ and attains the value φ_+ at a sufficiently small distance from the free boundary on the side of liquid phase and attains the value φ_- at a sufficiently small distance from the free boundary on the side of the solid phase. We assume that φ can approximately be represented as $\varphi = \phi((r - v_\mathbf{n}t)/\varepsilon)$, where $v_\mathbf{n}$ is the normal velocity of the free boundary $\Gamma(t)$. We put

$$\varepsilon^2 = \xi^2 a; \quad \alpha = \text{const}; \quad \xi, a \to 0; \quad \rho \equiv (r - v_\mathbf{n}t)/\varepsilon. \tag{3.19}$$

With regard to the last conditions, we use (3.17) to rewrite Eq. (3.16) as[1]

$$-\alpha v_{\mathbf{n}}\varepsilon\phi_\rho \cong \phi_{\rho\rho} + \varepsilon\mathcal{K}\phi_\rho + \cdots + \left(\phi - \phi^3\right)/2 + 2a\theta, \tag{3.20}$$

where the terms of the order of ε^2 are omitted. If there exists an expansion of the form

$$\phi = \phi^0 + \varepsilon\phi^1 + \cdots,$$

then (3.20) implies that the terms of the order $O(1)$ form the equation

$$\phi^0_{\rho\rho} + \frac{1}{2}\left[\phi^0 - \left(\phi^0\right)^3\right] = 0. \tag{3.21}$$

The last equation has solution

$$\phi^0(\rho) = \tanh(\rho/2). \tag{3.22}$$

Taking Eq. (3.21) (of the order $O(1)$) into account in Eq. (3.20), we can obtain an equation of the order $O(\varepsilon)$ (with regard to the fact that $a\varepsilon^{-1} = O(1)$ or less):

$$L\phi^1 = \phi^1_{\rho\rho} + \frac{1}{2}\left[1 - 3\left(\phi^0\right)^2\right]\phi^1 \cong \varepsilon\left(-\alpha v_{\mathbf{n}}\phi^0_\rho - \mathcal{K}\phi^0_\rho - \frac{2a}{\varepsilon}\theta\right) \equiv F. \tag{3.23}$$

Since the derivatives of the solution of the order $O(1)$ satisfy the homogeneous heat equation, for Eq. (3.23) and $L\phi^0_\rho = 0$, we can obtain the following solvability condition (for the rigorous derivation of these equations, see Sect. 3.4):

$$0 = \left(F, \phi^0_\rho\right) = \varepsilon \int_{-\infty}^{+\infty} \phi^0_\rho\left(-\alpha v_{\mathbf{n}}\phi^0_\rho - \mathcal{K}\phi^0_\rho - 2(a/\varepsilon)\theta\right) d\rho. \tag{3.24}$$

Since

$$\int_{-\infty}^{+\infty} \phi^0_\rho \, d\rho = 2,$$

condition (3.24) implies the identity

$$4\theta(x, t) \cong -\frac{\varepsilon}{a}\sigma_0\mathcal{K}(x, t) - \frac{\varepsilon}{a}\sigma_0 v_{\mathbf{n}}(x, t) \tag{3.25}$$

on $\Gamma(t)$, where

---

[1] For the rigorous asymptotic analysis of the phase field system, see in the next sections.

$$\sigma_0 \equiv \int\limits_{-\infty}^{+\infty} \left(\phi_\rho^0\right)^2 \mathrm{d}\rho = \frac{2}{3}.$$

Therefore, the quantity $\varepsilon a^{-1} = \xi a^{-1/2}$ is an important parameter.

Now we perform calculations to understand the value of $\varepsilon a^{-1}\sigma_0$ in Eq. (3.25). We solve Eq. (3.16) starting from the fact that the following relation must be satisfied:

$$\tau\varphi_t = \frac{\delta\mathcal{E}}{\delta\varphi},$$

where the free energy $\mathcal{E}$ is given by the equation

$$\mathcal{E}\{\varphi\} = \int\limits_{\Omega} \left(\frac{\xi^2}{2}(\nabla\varphi)^2 + \frac{1}{8a}\left(\varphi^2 - 1\right)^2 - 2\theta\varphi\right) \mathrm{d}x$$

for the function $g(\varphi)$ considered above. Here Ω is the spatial domain, where Eqs. (3.15) and (3.16) are considered, see Sect. 3.1. The surface tension σ is determined by the formula

$$\sigma \equiv \frac{\mathcal{E}\{\varphi\} - \frac{1}{2}\mathcal{E}\{\varphi_+\} - \frac{1}{2}\mathcal{E}\{\varphi_-\}}{A} \cong \frac{\mathcal{E}\{\phi^0\}}{A}, \tag{3.26}$$

where A is the area of the interface $\Gamma(t)$, see [6]. To calculate the leading terms of expansion (3.26), we multiply (3.21) by φ_r^0, integrate, and obtain

$$0 = \int\limits_{-\infty}^{r} \left\{\xi^2\varphi_r^0\varphi_{rr}^0 + \frac{1}{2}\varphi_r^0\left[\varphi^0 - \left(\varphi^0\right)^3\right]\right\} \mathrm{d}r, \tag{3.27}$$

whence we have

$$\frac{\xi^0}{2}(\varphi_r^0)^2 = \frac{1}{8}(\varphi^2 - 1). \tag{3.28}$$

Therefore, the free energy in formula (3.26) can be written as

$$\sigma \cong \frac{\mathcal{E}\{\varphi^0\}}{A} = \int\limits_{-\infty}^{\infty} \xi^2(\varphi_r^0)^2 \, \mathrm{d}r = \frac{\varepsilon}{a}\int\limits_{-\infty}^{\infty} \left(\phi_\rho^0\right)^2 \mathrm{d}\rho. \tag{3.29}$$

We note that the difference between the densities of entropies of the liquid and solid phases Δs is given by the formula [8]

$$\Delta s \equiv \frac{-\dfrac{\partial \mathcal{E}}{\partial u}\{\varphi_+\} + \dfrac{\partial \mathcal{E}}{\partial u}\{\varphi_-\}}{V} \cong 4, \tag{3.30}$$

where V is the volume.

The condition on the free boundary for modified Stefan problem (3.9) is obtained as a quantity of the order $O(1)$ in the framework of our heuristic calculations provided that

$$\varepsilon a^{-1} = \xi a^{-1/2} = O(1)$$

or less. But if $\xi a^{-1/2} \to 0$, then we obtain the classical Stefan condition

$$\theta(x, t) = 0, \quad x \in \Gamma(t). \tag{3.31}$$

Finally, if $\alpha \to 0$ while $\xi a^{-1/2} = O(1)$, then we obtain condition (3.8) in the limit.

In all cases of the choice of parameters considered above, the width of the interface is of the order of ε and the solution φ is approximated by formula (3.21). Thus, at a sufficiently large distance from the interface, the function φ is equal to a constant (with an arbitrary accuracy with respect to ε) so that the heat condition (3.4) is satisfied. Condition (3.5) for the latent heat of melting l can be obtained as the result of integration as $\varepsilon \to 0$. The process of origination of conditions on the free boundary will be discussed in detail in Sect. 3.4.

It is clear that, in addition to the three limits of Stefan problem type, we obtain principally different properties when considering the values $\varepsilon = \xi a^{1/2}$ and $\varepsilon a^{-1} = \xi a^{-1/2}$, i.e., when considering the role played by the interface thickness and the surface tension. In the careful consideration of the physical processes, we must compare the molecular and atomic forces represented by the parameter ξ and the processes at a deeper level determined by the parameter a^{-1}. The latter are completely described by the potential $a^{-1} g(\varphi)$. This potential can be interpreted as the energy barrier between two phases, which, of course, depends on a specific microscopic consideration. The double potential well $-(\varphi^2 - 1)^2/8a$ can also be represented [31] from the standpoint of the ϕ^4-field theory as the probability measure which expresses the "preference" of the transition to the solid, liquid, or intermediate state on the phase diagram.

3.3   Asymptotic Solution of the Phase Field System and Modified Stefan Problem

In this section, based on [20], we show how to construct an asymptotic (with respect to the parameter $\varepsilon \ll 1$) solution of the system of phase field equations such that the calculation of its uniform limit as $\varepsilon \to 0$ permits obtaining well-posed conditions on the free boundary, i.e., results in the modified Stefan problem with Gibbs–Thomson condition for the temperature. The approach described here (more rigorous than that in the preceding section) is close to that of Oleinik's small viscosity method [46], where the "correct" Rankine–Hugoniot conditions for the discontinuous solution of the Hopf equation (in our case, the conditions on the free boundary for the modified Stefan problem) were obtained as a result of considering the equation with a small viscosity (in our case, the phase field system) and subsequent passing to the limit as $\varepsilon \to 0$. The asymptotic solution is constructed in this section following the results obtained in [51, 53] by the modified two-scale method [19, 42, 43] for the solutions with localized "fast" variation.

For the presentation convenience and without loss of generality, we leave all connections with the physical problem and write the initial boundary-value problem for the phase field system in the form

$$\frac{\partial \theta}{\partial t} - \Delta \theta = -\frac{\partial \varphi}{\partial t}; \tag{3.32}$$

$$\varkappa \varepsilon^2 \frac{\partial \varphi}{\partial t} - \varepsilon^2 \Delta \varphi = \varphi - \varphi^3 + \varepsilon \varkappa_1 \theta; \tag{3.33}$$

$$\theta\big|_{t=0} = \theta^0(x, \varepsilon); \quad \varphi\big|_{t=0} = \varphi^0(x, \varepsilon); \tag{3.34}$$

$$k_1 \frac{\partial \theta}{\partial \mathbf{n}}\bigg|_{\Sigma} + k_2 \theta\big|_{\Sigma} = g^0(x, \varepsilon)\big|_{\Sigma}; \quad \varphi\big|_{\Sigma} = \sigma. \tag{3.35}$$

Here $x \in \Omega$, where $\Omega \subset \mathbb{R}^m$, $m \geq 1$, is a bounded domain with a smooth boundary $\partial \Omega$, $t \in [0, T]$, $\Sigma = \partial \Omega \times [0, T]$, $\varepsilon \ll 1$ is a small parameter, $\varkappa > 0$ and $\varkappa_1 > 0$ are constants, $g^0 \in C^\infty(\Sigma)$ is a given function, $\sigma = 1$ (or $\sigma = -1$), and $\partial/\partial \mathbf{n}$ is the outward normal derivative on $\partial \Omega$. In problem (3.32)–(3.35) it is necessary to find a pair of smooth (C^∞) functions $\theta > 0$, φ.

We assume that, for $t = 0$, a smooth closed nonsingular surface Γ_0 of dimension $m - 1$ is given in Ω, and $\Gamma_0 \cap \partial \Omega = \varnothing$. By Ω_0^- we denote the subdomain of Ω bounded by the surface Γ_0, and let $\Omega_0^+ = \Omega \setminus \overline{\Omega_0^-}$. Now we formulate assumptions about the initial functions $\theta^0 > 0$, φ^0.

Assumption 3.1 The pointwise limit of $\theta^0(x, \varepsilon)$ as $\varepsilon \to 0$ is a positive continuous function of $x \in \Omega$.

Assumption 3.2 The pointwise limit of $\varphi^0(x, \varepsilon)$ as $\varepsilon \to 0$ is equal to 1 for $x \in \Omega_0^+$ and to -1 for $x \in \Omega_0^-$ in the case $\sigma = 1$ (or in the case $\sigma = -1$: $\lim_{\varepsilon \to 0} \varphi^0(x, \varepsilon) = -1$ for $x \in \Omega_0^+$ and $\lim_{\varepsilon \to 0} \varphi^0(x, \varepsilon) = 1$ for $x \in \Omega_0^-$).

In what follows, for definiteness, we assume that $\sigma = 1$. Assumptions 3.1 and 3.2 are preliminary, and they are insufficient for obtaining well-posed boundary conditions on the free boundary for the limit system of Eqs. (3.32), (3.33). To specify the conditions for θ and φ for $t = 0$ and for $t > 0$ is, in fact, the main goal in this section.

Remark 3.1 We show below that, as in the classical Stefan problem, choosing $\theta^0\big|_{\Gamma_0} = 0$ for the surface Γ_0 with zero mean curvature, we stabilize the free boundary and thus obtain an intermediate region.

Now we formulate the main result in this section, i.e., the modified Stefan problem. For this, we let Γ_t denote the $(m-1)$-dimensional smooth nonsingular surface determining the position of the interface between phases for each fixed $t \in [0, T]$. Let Ω_t^- be a subdomain bounded by Γ_t, and let $\Omega_t^+ = \Omega \setminus \overline{\Omega_t^-}$. We introduce the functions $\theta^\pm = \theta^\pm(x, t)$ and $\psi = \psi(x)$ as the solution of the problem

$$\frac{\partial \theta^\pm}{\partial t} = \Delta \theta^\pm, \quad (x, t) \in \Omega_t^\pm, \tag{3.36}$$

$$\theta^\pm\big|_{t=0} = \theta_\pm^0(x), \quad x \in \Omega_0^\pm, \tag{3.37}$$

$$\theta^+\big|_{\Gamma_t^+} = \theta^-\big|_{\Gamma_t^-}, \tag{3.38}$$

$$\left(\partial \theta^+/\partial \mathbf{n}_t\right)\big|_{\Gamma_t^+} - \left(\partial \theta^-/\partial \mathbf{n}_t\right)\big|_{\Gamma_t^-} = 2v_\mathbf{n}, \tag{3.39}$$

$$k_1\left(\partial \theta^+/\partial \mathbf{n}\right)\big|_\Sigma + k_2 \theta^+\big|_\Sigma = g^0(x, t)\big|_\Sigma, \tag{3.40}$$

$$|\nabla \psi|\operatorname{div}\left(\nabla \psi/|\nabla \psi|\right) + \varkappa - \varkappa_2|\nabla \psi|\check{\theta}(x, \psi) = 0, \tag{3.41}$$

$$\psi\big|_{\Gamma_0} = 0, \tag{3.42}$$

where Γ_t is the surface defined by the equation $t = \psi(x)$, $f^\pm\big|_{\Gamma_t^\pm}$ is the limit of f as $x \to x_t \pm 0$, $x_t \in \Gamma_t$,

$$\check{\theta}(x, \psi) = \theta^+\big|_{\Gamma_t} = \theta^-\big|_{\Gamma_t}, \quad \theta_\pm^0 = \lim_{\varepsilon \to 0} \theta^0(x, \varepsilon), \quad x \in \Omega_0^\pm,$$

$\varkappa_2 = 3\varkappa_1/\sqrt{2}$, $\partial/\partial \mathbf{n}_t = \langle \nabla \psi/|\nabla \psi|, \nabla \rangle$ is the derivative along the normal to Γ_t, and $v_\mathbf{n}$ is the velocity of motion of Γ_t in the direction of the normal $\nabla \psi/|\nabla \psi|$.

We point out the following important fact. For the nonsingular surface Γ_0 with nonzero mean curvature, Eq. (3.41) is quasilinear parabolic, where the role of time is played by the parameter along the normal to Γ_t. Thus, problem (3.41), (3.42) is the Cauchy problem with initial conditions on the closed manifold Γ_0 which is necessarily solvable for $0 < \psi \le T$ for some $T > 0$. Equation (3.41) can be rewritten in the form that admits the following geometric meaning:

$$\check{\theta}\big|_{\Gamma_t} = -\mathcal{K}_t/\varkappa_2 - (\varkappa/\varkappa_2)v_\mathbf{n}, \tag{3.43}$$

where $\mathcal{K}_t$ is the mean curvature of Γ_t. But we prefer the form (3.41) thus stressing that this is the equation "determining" the position of the free boundary at time t and closing conditions (3.37)–(3.40).

The problems of solvability of the modified Stefan problem were discussed in [48, 52], and we do not consider them here. For us it is important to consider the condition that the initial and boundary conditions are consistent [48]; we formulate this condition as follows.

Assumption 3.3 The surface Γ_0 and the limit of the function $\theta^0(x, \varepsilon)$ as $\varepsilon \to 0$ determine the vector $\nabla \psi|_{\Gamma_0}$ which coincides with the corresponding vector determined by (3.41) and (3.42).

Theorem 3.1 *Assume that Assumptions 3.1–3.3 are satisfied and problem (3.36)–(3.42) has a smooth solution for $t \leq T$. Then for $t \in [0, T]$, with an accuracy of $O(\varepsilon)$, there exists an asymptotic solution $\theta(x, t, \varepsilon)$, $\varphi(x, t, \varepsilon)$ of problem (3.32)–(3.35) such that, for $(x, t) \in \Omega_t^\pm$, $\lim\limits_{\varepsilon \to 0} \theta = \theta^\pm$ and $\lim\limits_{\varepsilon \to 0} \varphi = \pm 1$ in the case $\sigma = 1$ or $\lim\limits_{\varepsilon \to 0} \varphi = \mp 1$ in the case $\sigma = -1$.*

In what follows, we not only prove Theorem 3.1 but also structurally construct the formal asymptotic solution with an accuracy of $O(\varepsilon^2)$ and present a method for constructing an asymptotics with an arbitrary accuracy.

3.3.1 Construction of an Asymptotic Solution

We need the following two classes of functions. We write

$$\mathcal{F} = \Big\{ f(\tau, x, t) \in C^\infty(\mathbb{R}^1 \times \Omega \times [0, T]), \exists\, f^\pm(x, t) \in C^\infty(\Omega \times [0, T]),$$

$$\lim_{\tau \to \pm\infty} \tau^n \frac{\partial^r}{\partial \tau^r} \frac{\partial^{|\alpha|}}{\partial x^{|\alpha|}} \frac{\partial^{|\gamma|}}{\partial t^{|\gamma|}} \big(f(\tau, x, t) - f^\pm(x, t) \big) = 0$$

$$\forall n \geq 0,\ r \geq 0,\ \gamma \geq 0,\ |\alpha| \geq 0 \Big\},$$

$$\mathbf{S} = \big\{ f(\tau, x, t) \in \mathcal{F},\ f^+ = f^- = 0 \big\}.$$

Lemma 3.1 *1. For any functions $S(x, t) \in C^\infty(\Omega \times [0, T])$ such that $(\partial S/\partial t)|_{\Gamma_t} \neq 0$, where $\Gamma_t = \big\{ (x, t) \in \Omega \times [0, T],\ S(x, t) = 0 \big\}$, and any functions $f(\tau, x, t) \in \mathcal{F}$, the following relation holds:*

$$f\left(\frac{S(x, t)}{\varepsilon}, x, t \right) = f\left(\frac{\beta(t - \psi(x))}{\varepsilon}, x, t \right) + O(\varepsilon),$$

where $t = \psi(x)$ is the equation of the surface $S(x, t) = 0$, $\beta = (\partial S/\partial t)|_{\Gamma_t}$.

2. Let $\chi(\tau, x, t)$ be a function of class $\mathcal{F}$ such that $\chi^\pm = \pm 1$. Then for any function $f \in \mathcal{F}$, we have the formula

$$f = \frac{f^+ + f^-}{2} + \frac{f^+ - f^-}{2} \chi(\tau, x, t) + \omega(\tau, x, t), \tag{3.44}$$

where ω is a function from **S**.

3. *For any functions* $f(\tau, x, t) \in$ **S** *and* $g^0(x, t) \in C^\infty$, *the following relations hold:*

$$(t - \psi)^k f\left(\frac{t - \psi}{\varepsilon}, x, t\right) = O(\varepsilon^k), \qquad k \geq 0,$$

$$g^0(x, t) f\left(\frac{t - \psi}{\varepsilon}, x, t\right) = g^0(x, \psi) f\left(\frac{t - \psi}{\varepsilon}, x, \psi\right) + O(\varepsilon).$$

The proof of the lemma, which is quite obvious, is given in [42].

Let us construct a self-similar asymptotic solution of problem (3.32)–(3.35). First, we note that, by Assumptions 3.1 and 3.2, the limit as $\varepsilon \to 0$ of the leading term of the asymptotics of φ is discontinuous on Γ_t, i.e., of the type of Heaviside function, the limit of the leading term of the asymptotics of θ is a continuous function and, simultaneously, its derivative along the normal to Γ_t is in general strongly discontinuous on Γ_t. Thus, Assumptions 3.1 and 3.2 determine the form of the proposed asymptotic solution

$$\theta(x, t, \varepsilon) = \Xi^N(x, t, \varepsilon) + \mathcal{V}^N\left(\frac{S(x, t)}{\varepsilon}, x, t, \varepsilon\right) + O(\varepsilon^{N+1}), \tag{3.45}$$

$$\varphi(x, t, \varepsilon) = \mathcal{W}^N\left(\frac{S(x, t)}{\varepsilon}, x, t, \varepsilon\right) + \varepsilon \Phi^N(x, t, \varepsilon) + O(\varepsilon^{N+1}), \tag{3.46}$$

where

$$\mathcal{W}^N(\tau, x, t, \varepsilon) = \chi(\tau, x, t) + \sum_{j=1}^{N} \varepsilon^j W_j(\tau, x, t),$$

$$\Phi^N(x, t, \varepsilon) = \sum_{j=1}^{N} \varepsilon^{j-1} \varphi_j(x, t); \quad \Xi^N(x, t, \varepsilon) = \sum_{j=0}^{N} \varepsilon^j \theta_j(x, t),$$

$$\mathcal{V}^N(\tau, x, t, \varepsilon) = \rho(x, t) V_0(\tau, x, t) + \sum_{j=1}^{N} \varepsilon^j \big(\rho_j(x, t) V_j(\tau, x, t) + U_j(\tau, x, t)\big).$$

Here the unknown functions are

$$S, \ \rho, \ \rho_j, \ \theta_j, \ \varphi_j \in C^\infty\big(\Omega \times [0, T]\big),$$
$$V_j(\tau, x, t), \ U_j(\tau, x, t), \ W_j(\tau, x, t), \ \chi(\tau, x, t) \in \mathcal{F},$$

and

$$\chi^\pm = 1; \quad \left.\frac{\partial S}{\partial t}\right|_{\Gamma_t} \neq 0, \quad \rho\big|_{\Gamma_t} = 0; \quad \rho_j\big|_{\Gamma_t} = 0, \tag{3.47}$$

where $\Gamma_t = \{t = \psi(x), \ S(x, \psi(x)) = 0\}$.

By Lemma 3.1 and (3.47), without loss of generality, we assume

$$S = t - \psi(x); \quad V_j = \alpha_j^+ + \alpha_j^- \chi(\tau, x, t), \tag{3.48}$$

where $j = 0, 1, \ldots, N$,

$$\alpha_j^+ = \left(V_j^+(x, t) + V_j^-(x, t)\right)/2; \quad \alpha_j^- = \left(V_j^+(x, t) - V_j^-(x, t)\right)/2.$$

We substitute expansions (3.45), (3.46) into (3.32) and (3.33). Multiplying equation (3.32) by ε^2, we obtain

$$
|\nabla S|^2 \frac{\partial^2 \mathcal{V}^N}{\partial \tau^2} - \varepsilon \left(\frac{\partial \mathcal{V}^N}{\partial \tau} + \frac{\partial \mathcal{W}^N}{\partial \tau} \right)
$$
$$
+ \varepsilon \{ 2\langle \nabla S, \nabla \rangle + \Delta S \} \frac{\partial \mathcal{V}^N}{\partial \tau} + \varepsilon^2 \Delta \left(\Xi^N + \mathcal{V}^N \right)
$$
$$
- \varepsilon^2 \frac{\partial}{\partial t} \left(\Xi^N + \mathcal{V}^N + \mathcal{W}^N + \varepsilon \Phi^N \right)\big|_{\tau = S/\varepsilon} = O\left(\varepsilon^{N+3}\right), \tag{3.49}
$$

and from (3.33) we derive

$$
\mathcal{W}^N - \left(\mathcal{W}^N\right)^3 + |\nabla S|^2 \frac{\partial^2 \mathcal{W}^N}{\partial \tau^2} + \varepsilon \left\{ \varkappa_1 \left(\Xi^N + \mathcal{V}^N \right) \right.
$$
$$
+ \left(2 \langle \nabla S, \nabla \rangle + \Delta S \right) \frac{\partial \mathcal{W}^N}{\partial \tau} + \Phi^N \left(1 - 3 \left(\mathcal{W}^N \right)^2 \right)
$$
$$
\left. - \varkappa \frac{\partial \mathcal{W}^N}{\partial \tau} \right\} + \varepsilon^2 \left\{ \left(\Delta - \varkappa \frac{\partial}{\partial t} \right) \mathcal{W}^N - 3 \left(\Phi^N \right)^2 \mathcal{W}^N \right\}
$$
$$
+ \varepsilon^3 \left\{ \left(\Delta - \varkappa \frac{\partial}{\partial t} \right) \Phi^N - \left(\Phi^N \right)^3 \right\} \Bigg|_{\tau = S/\varepsilon} = O\left(\varepsilon^{N+1}\right). \tag{3.50}
$$

First, we determine regular terms of expansions (3.45) and (3.46). Passing to the limit as $\tau \to \pm\infty$ in (3.49) and (3.50), we obtain

$$
\Delta \left(\Xi^N + \left(\mathcal{V}^N\right)^\pm \right) - \frac{\partial}{\partial t} \left(\Xi^N + \left(\mathcal{V}^N\right)^\pm + \left(\mathcal{W}^N\right)^\pm + \varepsilon \Phi^N \right) = O\left(\varepsilon^{N+1}\right),
$$

$$
\left(\mathcal{W}^N\right)^\pm - \left(\left(\mathcal{W}^N\right)^\pm \right)^3 + \varepsilon \left\{ \Phi^N \left(1 - 3 \left(\left(\mathcal{W}^N\right)^\pm \right)^2 \right) \right.
$$
$$
\left. + \varkappa_1 \left(\Xi^N + \left(\mathcal{V}^N\right)^\pm \right) \right\} + \varepsilon^2 \left(\Delta - \varkappa \frac{\partial}{\partial t} - 3 \left(\Phi^N \right)^2 \right) \left(\mathcal{W}^N\right)^\pm
$$
$$
+ \varepsilon^3 \left(\Delta - \varkappa \frac{\partial}{\partial t} - \left(\Phi^N \right)^2 \right) \Phi^N = O\left(\varepsilon^{N+1}\right).
$$

From this, equating the terms of the same order in ε with zero, we obtain the relations $\chi^{\pm} = \pm 1$ (with regard to the initial conditions),

$$\mathcal{V}_0^{\pm} = \rho V_0^{\pm}, \quad \mathcal{V}_i^{\pm} = \rho V_i^{\pm} + U_i^{\pm}, \ i \geq 1,$$

the relations between φ_{j+1}, $W_{j+1}^{\pm}$ and θ_j:

$$\varphi_1 + W_1^{\pm} = \frac{\varkappa_1}{2}(\theta_0 + \mathcal{V}_0^{\pm}), \tag{3.51}$$

$$\varphi_{j+1} + W_{j+1}^{\pm} = \frac{\varkappa_1}{2}(\theta_j + \mathcal{V}_j^{\pm}) + f_j^{\pm}, \tag{3.52}$$

where $j = 1, 2, \ldots, N - 1$, and the equations

$$\left(\frac{\partial}{\partial t} - \Delta\right)(\theta_0 + \mathcal{V}_0^{\pm}) = 0, \tag{3.53}$$

$$\left(\frac{\partial}{\partial t} - \Delta\right)(\theta_j + \mathcal{V}_j^{\pm}) = -\frac{\partial}{\partial t}(\varphi_j + W_j^{\pm}), \tag{3.54}$$

where $j = 1, 2, \ldots, N - 1$ and $f_j^{\pm}$ are polynomials in $\varphi_i + W_i^{\pm}, i = 1, 2, \ldots, j$. In particular, $f_1^{\pm} = \mp(3/2)(\varphi_1 + W_1^{\pm})$. Now the left-hand sides of (3.49) and (3.50) are functions from **S**.

We construct $\mathcal{V}^N$ and $\mathcal{W}^N$ following the scheme developed in [19, 42, 43]; namely, for all terms in relations (3.49) and (3.50), we successively expand the quantities $O(\varepsilon^j), j = 0, 1, \ldots, N$, in the Taylor series with respect to t at the point $t = \psi(x)$ and use the relation

$$\tau = \frac{t - \psi(x)}{\varepsilon}.$$

Then we pass to functions with independent variables τ, x and determine them on the surface $t = \psi$. After this, we determine an infinitely differentiable continuation of these functions outside the surface Γ_t. We construct the continuation so as to ensure the existence of lower-order terms of the asymptotics with required properties.

Equating the terms of the order $O(1)$ in (3.50) and taking (3.47) into account, we obtain the following model equation for the function $\check{\chi} = \chi(\tau, x, \psi(x))$:

$$\check{\beta}^{-2}\frac{\partial^2 \check{\chi}}{\partial \tau^2} + \check{\chi} - \check{\chi}^3 = 0, \tag{3.55}$$

$$\check{\chi}\big|_{\tau \to \pm\infty} \to \pm 1,$$

where $\check{\beta} = 1/|\nabla S| = 1/|\nabla \psi|$, whence, on Γ_t, we have

$$\check{\chi}(\tau, x) = \tanh\left(\check{\beta}(\tau + s)/\sqrt{2}\right),$$

where $s = s(x) \in C^\infty(\Omega)$ is the "constant" of integration of model equation (3.55). For all $(x, t) \in \Omega \times [0, T]$, the function $\check{\chi}$ is continued identically, i.e., $\chi = \check{\chi}(\tau, x)$.

Further, we note that, by (3.47) and Lemma 3.1,

$$\frac{\partial \mathcal{V}^N}{\partial \tau} = O(\varepsilon), \quad \frac{\partial^2 \mathcal{V}^N}{\partial \tau^2} = O(\varepsilon).$$

Therefore, equating the terms of order $O(\varepsilon)$ with zero in (3.49) and (3.50) and using (3.51) and (3.53), we obtain

$$\check{\beta}^{-2}\frac{\partial^2 \check{U}_k}{\partial \tau^2} = G_k(\tau, x), \tag{3.56}$$

$$\hat{L}\omega_k = F_k(\tau, x). \tag{3.57}$$

Here $k = 1$,

$$\hat{L} = \check{\beta}^{-2}\frac{\partial^2}{\partial \tau^2} + 1 - 3\chi^2,$$

and for the functions U_j and $W_j \in \mathcal{F}$ we use a representation of the form (3.44), i.e., $U_j = \gamma_j^+ + \gamma_j^- \chi + u_j$, $W_j = \mu_j^+ + \mu_j^- \chi + \omega_j$, where $\gamma_j^\pm = (U_j^+ \pm U_j^-)/2$, $\mu_j^\pm = (W_j^+ \pm W_j^-)/2$, the functions $u_j = u_j(\tau, x) \in \mathbf{S}$ and $\omega_j = \omega_j(\tau, x) \in \mathbf{S}$, $j = 1, 2, \ldots, N$, and

$$G_1 = \frac{\partial \chi}{\partial \tau} + \left\{ 2\langle \nabla\psi, \nabla\rho\rangle \frac{\partial V_0}{\partial \tau} - \rho_t \check{\beta}^{-2}\tau\frac{\partial^2 V_0}{\partial \tau^2} \right\}\Bigg|_{t=\psi},$$

$$F_1 = \left(2\langle\nabla\psi, \nabla\rangle + \Delta\psi + \varkappa\right)\frac{\partial \chi}{\partial \tau} - \frac{3}{2}\varkappa_1\check{\theta}_0\left(1 - \chi^2\right),$$

where we write $\check{f} = f(\iota, x, t)\big|_{t=\psi(x)}$.

The following assertion holds (also see [42]).

Lemma 3.2 *For the solvability of Eq. (3.56) in $\mathcal{F}$ and Eq. (3.57) in $\mathbf{S}$, it is necessary and sufficient to satisfy the conditions*

$$G_k \in \mathbf{S}, \quad F_k \in \mathbf{S}, \tag{3.58}$$

$$\int_{-\infty}^{+\infty} G_k(\tau, x)\,d\tau = 0, \tag{3.59}$$

$$\int_{-\infty}^{+\infty} F_k(\tau, x)\frac{\partial \chi}{\partial \tau}\,d\tau = 0. \tag{3.60}$$

Conditions (3.58) for $k = 1$ are satisfied by (3.51) and (3.53), and condition (3.59) for $k = 1$ implies

$$\alpha_0^- \left(\rho_t \check{\beta}^{-2} + 2\langle \nabla\psi, \nabla\rho \rangle \right)\big|_{\Gamma_t} + 1 = 0. \tag{3.61}$$

We note that, for the functions ρ vanishing on the surface Γ_t,

$$\nabla\rho\big|_{\Gamma_t} = -\rho_t \nabla\psi\big|_{\Gamma_t}. \tag{3.62}$$

Therefore, we can rewrite relation (3.61) as

$$\rho_t |\nabla\psi|^2 \alpha_0^- \big|_{\Gamma_t} = 1. \tag{3.63}$$

We note that this implies the condition $\rho_t\big|_{\Gamma_t} \neq 0$, i.e., the normal derivative of the limit solution is necessarily discontinuous on Γ_t. The following expression is equivalent to (3.63):

$$\left\langle \frac{\nabla\psi}{|\nabla\psi|}, \nabla \right\rangle \rho \left(V_0^+ - V_0^- \right)\bigg|_{\Gamma_t} = \frac{-2}{|\nabla\psi|}. \tag{3.64}$$

By $\theta^\pm(x,t)$ we denote the pointwise limit of $\theta(x,t,\varepsilon)$ in $\Omega_t^\pm$ as $\varepsilon \to 0$ and, taking the continuity of the background $\theta_0(x,t)$ into account, transform condition (3.64) to the form (3.39).

Further, after simple calculations, considering condition (3.60) for $k = 1$, we obtain the following assertion.

Lemma 3.3 *Condition* (3.60) *for* $k = 1$ *is equivalent to the relation*

$$\check{\beta}^{-1}\mathrm{div}\,(\check{\beta}\nabla\psi) + \varkappa - (3/\sqrt{2})\check{\beta}^{-1}\varkappa_1\check{\theta}_0 = 0. \tag{3.65}$$

Since $\rho\big|_{\Gamma_t} = 0$, we have $\theta^+\big|_{\Gamma_t} = \theta^-\big|_{\Gamma_t} = \check{\theta}_0$ and (3.65) implies Eq. (3.41). Now we can integrate equations (3.56) and (3.57) for $k = 1$ and obtain

$$\check{U}_1 = \check{\gamma}_1^+(x) + \check{\gamma}_1^-(x)\chi + u_1(\tau, x); \quad \omega_1 = \omega_{11}(\tau, x) + s_1\chi_\tau,$$

where $\omega_{11} \in \mathbf{S}$ is a partial solution of (3.57), $\check{\gamma}_1^+$ and s_1 are arbitrary functions, and

$$\check{\gamma}_1^- = \frac{s}{|\nabla\psi|^2}, \quad u_1 = \frac{\sqrt{2}}{\check{\beta}}\left(\xi\chi - \ln\mathrm{ch}\,\xi \right)\bigg|_{\xi=\check{\beta}(\tau+s)/\sqrt{2}} \in \mathbf{S}. \tag{3.66}$$

We continue $\mathcal{V}_0^N = \rho V_0$ outside the surface Γ_t. For this, we let $\theta^\pm = \theta^\pm(x,t)$ denote the solution of problem (3.36)–(3.42) in $\Omega_t^\pm$, and let $\tilde{\theta}^\pm = \tilde{\theta}^\pm(x,t)$ be respective smooth continuations of $\theta^\pm$ into the domains $\Omega_t^\mp$ such that $\tilde{\theta}^\pm$ also satisfy the heat conduction equation. Then we set

$$\rho\alpha_0^- = (\tilde{\theta}^+ - \tilde{\theta}^-)/2, \quad \mu_1^- = \varkappa_1\rho\alpha_0^-/2, \quad \theta_0 = \rho\alpha_0^+ = (\tilde{\theta}^+ + \tilde{\theta}^-)/2,$$

$$\varphi_1 + \mu_1^+ = \varkappa_1(\theta_0 + \rho\alpha_0^+)/2,$$

and define the leading term of the asymptotics of θ in Ω as

$$\theta(x,t,\varepsilon) = \frac{\tilde{\theta}^+ + \tilde{\theta}^-}{2} + \frac{\tilde{\theta}^+ - \tilde{\theta}^-}{2}\chi(\tau,x) + O(\varepsilon), \tag{3.67}$$

and the first correction in the asymptotic expansion of φ as

$$\varphi(x,t,\varepsilon) = \chi(\tau) + \varepsilon\Big\{(\varkappa_1/4)(\tilde{\theta}^+ + \tilde{\theta}^-) + (\varkappa_1/4)$$
$$\times\,(\tilde{\theta}^+ - \tilde{\theta}^-)\chi(\tau,x) + \omega_1(\tau,x)\Big\} + O(\varepsilon^2), \tag{3.68}$$

where $\tau = \check{\beta}(\sqrt{2})^{-1}(t - \psi + \varepsilon s)/\varepsilon$.

We consider the terms of order $O(\varepsilon^2)$ in (3.49) and (3.50). Using formulas (3.52) and (3.54) for $j = 1$, we obtain Eqs. (3.56) and (3.57) for $k = 2$, where

$$G_2 = \alpha_1^-\left(|\nabla\psi|^2\rho_{1t}\tau\chi_{\tau\tau} - 2\langle\nabla\psi, \nabla\rho_1\rangle\chi_\tau\right)\Big|_{\Gamma_t} + s_1\chi_{\tau\tau} + g_2(\tau,x)$$

and g_2, F_2 are some functions of ω_{11}, u_1, τ, χ, and $\tilde{\theta}^\pm$ for $t = \psi(x)$.

Similarly, we verify that condition (3.58) for $k = 2$ is satisfied by (3.52) and (3.54) for $j = 1$, and condition (3.59) for $k = 2$ implies

$$\rho_{1t}\alpha_1^-|\nabla\psi|^2\Big|_{\Gamma_t} = -\frac{1}{2}\int\limits_{-\infty}^{+\infty} g_2(\tau,x)\,d\tau. \tag{3.69}$$

By $\theta_1^\pm = \theta_1 + \rho_1 V_1^\pm + U_1^\pm$ we denote the limit value as $\tau \to \pm\infty$ of the first correction in the asymptotic expansion of $\theta(x,t,\varepsilon)$ and note that we have already obtained formula (3.66) for $\check{\gamma}_1^- = (U_1^+ - U_1^-)/2\big|_{\Gamma_t}$. Then using (3.63) and a relation similar to (3.62), we can transform condition (3.69) to the form similar to (3.39):

$$\frac{\partial}{\partial\mathbf{n}_t}\theta_1^+\Big|_{\Gamma_t^+} - \frac{\partial}{\partial\mathbf{n}_t}\theta_1^-\Big|_{\Gamma_t^-} = -p_1 v_\mathbf{n}, \tag{3.70}$$

where $p_1 = \int\limits_{-\infty}^{+\infty} g_2(\tau,x)\,d\tau + 2\langle\nabla\psi, \nabla\rangle\left(s/|\nabla\psi|^2\right)$.

Further, using simple but cumbersome computations, we can reduce condition (3.60) for $k = 2$ to the linear nonhomogeneous equation

$$\hat{\mathcal{L}}s = f(x), \tag{3.71}$$

where $\hat{\mathcal{L}}$ is a variation in the operator in (3.65) and the right-hand side f can be calculated by using the functions defined in (3.67) and (3.68). Supplementing Eq. (3.71) with the initial condition

$$s\big|_{\Gamma_0} = s^0(x)\big|_{\Gamma_0} \tag{3.72}$$

with a certain function $s^0 \in C^\infty(\Gamma_0)$, we obtain the Cauchy problem for determining the first correction $s(x)$ to the phase ψ which is solvable in the same neighborhood of the surface Γ_0 as problem (3.41), (3.42).

Now we can integrate Eqs. (3.56) and (3.57) for $k = 2$ and obtain:

$$\check{U}_2 = \check{\gamma}_2^{\,+}(x) + \check{\gamma}_2^{\,-}(x)\chi + u_2(\tau, x),$$
$$\omega_2 = \omega_{21}(\tau, x) + s_2(x)\chi_\tau,$$

where $u_2 \in \mathbf{S}$, $\omega_{21} \in \mathbf{S}$, $\check{\gamma}_2^{\,-} = s\rho_1\alpha_1^-\big|_{\Gamma_t} - s_1 + f_2^-(x)$, and $\check{\gamma}_2^{\,+}$ and s_2 are some still arbitrary functions.

We define the continuation $(\mathcal{V}_1^N - \mathcal{V}_0^N)/\varepsilon = \rho V_1 + U_1$ outside the surface Γ_t. For this, we let $\theta_1^\pm = \theta_1^\pm(x, t)$ denote the solution of the problem

$$\frac{\partial \theta_1^\pm}{\partial t} - \Delta\theta_1^\pm = -\frac{\varkappa_1}{2}\frac{\partial \theta^\pm}{\partial t}, \quad (x, t) \in \Omega_t^\pm, \tag{3.73}$$

$$\theta_1^\pm\big|_{t=0} = \theta_\pm^1(x), \quad x \in \Omega_0^\pm, \tag{3.74}$$

$$\theta_1^+\big|_{\Gamma_t} = \theta_1^-\big|_{\Gamma_t}, \tag{3.75}$$

$$\frac{\partial \theta_1^+}{\partial \mathbf{n}_t}\bigg|_{\Gamma_t^+} - \frac{\partial \theta_1^-}{\partial \mathbf{n}_t}\bigg|_{\Gamma_t^-} = -p_1 v_{\mathbf{n}}, \tag{3.76}$$

$$k_1\frac{\partial \theta_1^+}{\partial \mathbf{n}_t} + k_2\theta_1^+\bigg|_{\Sigma} = 0, \tag{3.77}$$

where $\theta_\pm^1(x) = \lim_{\varepsilon \to 0}\big(\theta^0(x, \varepsilon) - \theta_\pm^0(x)\big)/\varepsilon,\, x \in \Omega_0^\pm$.

Let $\tilde{\theta}_1^\pm(x, t)$ be respective smooth continuations of $\theta_1^\pm$ in the domain $\Omega_t^\pm$ such that each of the functions $\tilde{\theta}_1^\pm$ satisfies the nonhomogeneous heat equation with the right-hand side

$$\tilde{f}_1^\pm = -\frac{\varkappa_1}{2}\frac{\partial \tilde{\theta}^\pm}{\partial t}.$$

We put

$$\rho_1\alpha_1^- + \gamma_1^- = (\tilde{\theta}_1^+ - \tilde{\theta}_1^-)/2,$$
$$\theta_1 + \rho_1\alpha_1^+ + \gamma_1^+ = (\tilde{\theta}_1^+ + \tilde{\theta}_1^-)/2,$$
$$\mu_2^- = \frac{\varkappa_1}{4}(\tilde{\theta}_1^+ - \tilde{\theta}_1^-) - \frac{3}{16}\varkappa_1^2\big((\tilde{\theta}^+)^2 + (\tilde{\theta}^-)^2\big),$$
$$\varphi_2 + \mu_2^+ = \frac{\varkappa_1}{4}(\tilde{\theta}_1^+ + \tilde{\theta}_1^-) - \frac{3}{16}\varkappa_1^2\big((\tilde{\theta}^+)^2 - (\tilde{\theta}^-)^2\big)$$

and determine the asymptotics of θ with an accuracy of $O(\varepsilon^2)$ in Ω:

$$\theta(x,t,\varepsilon) = \frac{\tilde{\theta}^+ + \tilde{\theta}^-}{2} + \frac{\tilde{\theta}^+ - \tilde{\theta}^-}{2}\chi$$
$$+ \varepsilon\left\{\frac{\tilde{\theta}_1^+ + \tilde{\theta}_1^-}{2} + \frac{\tilde{\theta}_1^+ - \tilde{\theta}_1^-}{2}\chi + u_1(\tau,x)\right\} + O(\varepsilon^2) \qquad (3.78)$$

and the asymptotics of φ with an accuracy of $O(\varepsilon^3)$:

$$\varphi(x,t,\varepsilon) = \chi + \varepsilon\left\{\frac{\varkappa_1}{4}(\tilde{\theta}^+ + \tilde{\theta}^-) + \frac{\varkappa_1}{4}(\tilde{\theta}^+ - \tilde{\theta}^-)\chi + \omega_1\right\}$$
$$+ \varepsilon^2\left\{\varphi_2 + \mu_2^+ + \mu_2^-\chi + \omega_2(\tau,x)\right\} + O(\varepsilon^3). \qquad (3.79)$$

Thus, calculating the lower-order term of the asymptotic expansion of θ and φ can be reduced to solving modified Stefan problem (3.71), (3.73)–(3.77) and solving ordinary equations (3.56) and (3.57) for $k = 2$. In this case, Eq. (3.71) is the Gibbs–Thomson correction condition and can be used to specify the position of the interface between the phases with an accuracy of $O(\varepsilon^2)$.

For the solvability of problem (3.71), (3.73)–(3.77), we must impose the following conditions on the initial functions $\theta_\pm^1(x)$, $x \in \Omega_0^\pm$, $\varphi_\pm^1(x) = \lim_{\varepsilon\to 0}(\varphi^0(x,\varepsilon) \mp 1)/\varepsilon$, $x \in \Omega_0^\pm$, and $s^0(x)\big|_{\Gamma_0}$ which are similar to the previous conditions, namely, to Assumptions 3.1, 3.2, and 3.3.

Assumption 3.4 The functions $\theta_\pm^1(x)$ belong to $C(\Omega)$.

Assumption 3.5 The functions $\varphi_\pm^1$ are equal to $(\varkappa_1/2)\theta_\pm^0(x)$, $x \in \Omega_0^\pm$.

Assumption 3.6 The initial data $s^0(x)$ and $\theta_\pm^1(x)$ are consistent on Γ_0.

The asymptotic solution (3.78), (3.79) is self-similar.

Correspondingly, the initial values (3.34) θ^0, φ^0 must be special and satisfy the following condition.

Assumption 3.7 The functions θ^0 and φ^0 satisfy the relations

$$\theta^0(x,\varepsilon) - \theta(x,t,\varepsilon)\big|_{t=0} = O(\varepsilon^2),$$
$$\varphi^0(x,\varepsilon) - \varphi(x,t,\varepsilon)\big|_{t=0} = O(\varepsilon^3),$$

where $\theta(x,t,\varepsilon)$ and $\varphi(x,t,\varepsilon)$ are the functions defined in (3.78) and (3.79).

Theorem 3.2 *Assume that Assumptions 3.1–3.7 are satisfied and, for $t \leq T$, there exists a smooth solution of problem (3.36)–(3.42). Then, for $t \in [0,T]$, with an accuracy of $O(\varepsilon^2)$, there exists an asymptotic solution $\theta(x,t,\varepsilon)$, $\varphi(x,t,\varepsilon)$ of problem (3.32)–(3.35). For $(x,t) \in \Omega_t^\pm$,*

$$\lim_{\varepsilon\to 0}\theta = \theta^\pm, \quad \lim_{\varepsilon\to 0}\frac{\theta - \theta^\pm}{\varepsilon} = \theta_1^\pm,$$
$$\lim_{\varepsilon\to 0}\varphi = \pm 1, \quad \lim_{\varepsilon\to 0}\frac{\varphi \mp 1}{\varepsilon} = \frac{\varkappa_1}{2}\theta^\pm.$$

The subsequent terms of the asymptotic expansion are constructed similarly. We note that, at each step of the asymptotic procedure, there remain two undetermined "constants" of integration, i.e., $s_i(x)$ and $\breve{\gamma}_i^+(x)$, which are determined at the next step by using conditions (3.56), (3.57) for $k = i + 1$.

3.3.2 Examples

We consider several examples related to the two-dimensional case, $x \in \mathbb{R}^2$. By λ, ζ we denote the local coordinates in a neighborhood of the curve Γ_0 and assume that $\zeta \in [0, \zeta_0]$ is the coordinate on Γ_0. We note that condition (3.39) for $t = 0$ and the initial data determine the vector $\nabla \psi \big|_{\Gamma_0}$. We choose λ so that $\lambda > 0$ along $\nabla \psi \big|_{\Gamma_0}$, pass to the coordinates (λ, ζ) in Eq. (3.41), and perform the Mises transformation, i.e., pass from (λ, ζ) to the variables (ψ, ζ). We write

$$\lambda = R(\psi, \zeta),$$

to rewrite Eq. (3.41) as

$$\frac{\partial R}{\partial \psi} = a \frac{\partial^2 R}{\partial \zeta^2} + b \breve{\theta}\big(\omega, x(R, \zeta)\big) - F, \quad \psi > 0, \tag{3.80}$$

$$R\big|_{\psi=0} = 0, \quad \frac{\partial^k R}{\partial \zeta^k}\bigg|_{\zeta=0} = \frac{\partial^k R}{\partial \zeta^k}\bigg|_{\zeta=\zeta_0}, \tag{3.81}$$

where $k = 0, 1, \ldots, a = \varkappa^{-1}\{\lambda, \zeta\}^2 A^{-1}, \{\lambda, \zeta\} = \lambda_1 \zeta_2 - \lambda_2 \zeta_1 \big|_{x=x(R,\zeta)}, b = \varkappa_2 \sqrt{A}$, $A = |\nabla_x \zeta|^2 (R_\zeta)^2 - 2\langle \nabla_x \zeta, \nabla_x \lambda \rangle R_\zeta + |\nabla \lambda|^2 \big|_{x=x(R,\zeta)}$,

$$F = (\varkappa A)^{-1}\Big((\lambda_{11} - R_\zeta \zeta_{11}) \times (\lambda_2 - R_\zeta \zeta_2)^2$$

$$+ (\lambda_{22} - R_\zeta \zeta_{22})(\lambda_1 - R_\zeta \zeta_1)^2 - 2(\lambda_{12} - R_\zeta \zeta_{12})$$

$$\times (\lambda_1 - R_\zeta \zeta_1)(\lambda_2 - R_\zeta \zeta_2)\Big)\bigg|_{x=x(R,\zeta)},$$

$$f_i = \frac{\partial f}{\partial x_i}, \quad f_{ij} = \frac{\partial^2 f}{\partial x_i \partial x_j}, \quad R_\zeta = \frac{\partial R}{\partial \zeta}.$$

It is easy to see that $a > 0$ for the nonsingular curve Γ_0.

Example 3.1 Let Γ_0 be a circle of radius $c = $ const. After simple transformations, we obtain

$$a = a_1/c^2; \quad a_1 = (\varkappa A)^{-1}, \quad b = \varkappa_3 \sqrt{A}(1 - R/c)^{-1},$$
$$A = (R_\zeta/c)^2 + (1 - R/c)^2; \quad F = F_1/c, \quad \varkappa_3 = \varkappa_2/\varkappa, \tag{3.82}$$
$$F_1 = -\big(\varkappa A(1 - R/c)\big)^{-1}\big((1 - R/c)^2 + 2(R_\zeta/c)^2\big).$$

Further, to simplify the problem, we assume that the initial value θ in a neighborhood of Γ_0 is radially symmetric. Then, for sufficiently small t and ψ, we can assume that $\check{\theta} = \check{\theta}(\psi, \lambda)$ and $R = R(\psi)$. Correspondingly, in this case, (3.80) becomes the ordinary differential equation

$$\frac{dR}{d\psi} = \varkappa_3 \check{\theta}(\psi, R) + \frac{1}{\varkappa c}(1 - R/c)^{-1}. \tag{3.83}$$

Example 3.2 Let Γ_0 be a circle of radius $c = 1/\delta \gg 1$. Under the assumption that δ is a small parameter, we obtain the problems of constructing an asymptotics of the solution of the equation

$$\frac{\partial R}{\partial \psi} = \delta^2 a_1 \frac{\partial^2 R}{\partial \zeta^2} + b\check{\theta}\big(\psi, x(R, \zeta, \delta)\big) - \delta F_1. \tag{3.84}$$

Let $\check{\theta}$ be a slowly varying function of ζ, i.e.,

$$\check{\theta} = \check{\theta}_0(R, \psi, \zeta) + O(\delta).$$

Then the solution of Eq. (3.84) can be constructed by the regular perturbation theory. By setting $R = R_0(\psi, \zeta) + \delta R_1(\psi, \zeta) + \cdots$, we obtain in the leading term

$$\partial R_0/\partial \psi = \varkappa_3 \check{\theta}_0(R_0, \psi, \zeta). \tag{3.85}$$

Example 3.3 Let Γ_0 be the line $x_1 = 0$. We put $\lambda = x_1, \zeta = x_2$ to obtain $a = (\varkappa A)^{-1}$, $A = 1 + (R_\zeta)^2, b = \varkappa_3 \sqrt{A}, F = 0$.

Assume that, in a neighborhood of Γ_0, the function $\check{\theta}$ is independent of ζ. Then, in this neighborhood, $R = R(\psi)$ and Eq. (3.80) becomes

$$\frac{dR}{d\psi} = \varkappa_3 \check{\theta}(\psi, R). \tag{3.86}$$

Thus, Examples 3.1–3.3 show that the behavior of the boundary Γ_t is significantly different in the case of nonzero curvature Γ_0 and in the cases of small and zero curvature. In the first case, the motion of the boundary is determined by parabolic equations (3.80), (3.81), (3.82), and in the second case, the boundary moves with a greater velocity according to ordinary differential equations (3.85), (3.86). Further, in the classical Stefan problem, we choose $\check{\theta} = 0$, then in the case of small and zero curvature, we obtain $R_\psi = 0$. This means that the boundary is stabilized and, as a consequence, a transient region is formed [14]. In the case of finite curvature, it follows from Eq. (3.83) that no stabilization can occur.

3.4 Weak Solution of the Phase Field System and the Melting Zone Model

In this section, following [22], we discuss the problems related to the construction of a weak asymptotic solution of the phase field system and the condition arising on the interface between phases (Rankine–Hugoniot-type conditions). We also consider the problem of modeling the mushy region between the phase states in the material.

The last problem is very interesting from the mathematical viewpoint and is exotic from the physical viewpoint. It is interesting that a version of the mushy region problem is related to the problem of beginning of the cathode melting which is discussed in Chap. 4. The mushy region construction is based on the definition of generalized solution to phase-field system. It turned out that this definition (and, generally speaking, that for most of the nonlinear problems involving PDE's) should be constructed in a special nonclassical way. We discuss this matter below and widely use the definition constructed in Chap. 4. In this section, we consider the simplest form of this system

$$l\frac{\partial \varphi}{\partial t} + \frac{\partial \theta}{\partial t} = k\Delta\theta, \tag{3.87}$$

$$\tau\frac{\partial \varphi}{\partial t} = \xi^2\Delta\varphi + \frac{\varphi - \varphi^3}{a} + \kappa\theta. \tag{3.88}$$

Here θ is the temperature and φ is the order function. In this case, the value $\varphi = 1$ corresponds to the liquid state of matter, and $\varphi = -1$, to the solid state, l is a parameter (physically, it corresponds to the relaxation time of the system), and the coefficients τ, ξ, a are treated as small parameters, see Sect. 3.2.

In several years after the appearance of [6, 8], many works have been published, where passing to the limit was justified, the class of initial data for which one obtains one or another limiting problem was improved, the asymptotic expansions were justified, and questions related to some other aspects of such problems were discussed. We do not describe or compare these results but only refer to [3, 10, 26, 40, 41, 45, 48, 54], where problems most similar to our problem were considered. These works also contain an extensive bibliography related to the phase field system, the Allen–Cahn equation (the second equation in phase field system (3.88) for $\kappa = 0$), and related problems.

But in these studies, the main attention was paid to the problems, where the limit $\overline{\varphi^0}(x)$ of the initial order function $\varphi^0(x, \varepsilon)$ belongs to the space $BVC(\Omega)$. We use the standard notation BV for the space of functions of bounded variation and

$$BVC(\Omega) = \{\varphi, \ \varphi \in BV(\Omega), \ |\varphi| = 1\}.$$

Under such initial conditions and some additional assumptions, the limit of the order function $\varphi(x, t, \varepsilon)$ also belongs to the space $BVC(\Omega)$ for almost all $t \in [0, T]$. Correspondingly, the limiting problem has at most finitely many surfaces Γ_t of codimension 1 on which the temperature $\bar{\theta}$ has a weak discontinuity (i.e., $\bar{\theta} \in C$ and $\bar{\theta} \notin C^1$). The surfaces Γ_t separate the domains $\Omega_t^{\pm}$ filled with the matter in different aggregate states. In other words, we have a situation with a fast varying perturbation localized on the surface Γ_t.

But, there is another known situation in which, along with the domains $\Omega_t^{\pm}$ occupied by the solid and liquid phases, there also exists a domain Ω_t^* filled with the matter in the intermediate phase state (the so-called mushy region). In the corresponding literature, as far as we know, the problem with mushy region is considered from the standpoint of limit equations [17, 38, 44, 49].

Starting from the approximation properties of the phase field system, one can expect that the situation with mushy region can be described as passing to the limit in the phase field model with special initial data. To verify these hypotheses is the main goal in this section.

In Eqs. (3.87) and (3.88), we put $\tau = \xi^2 = a, \ l = 1$, and $k = 1$. We denote the small parameter by $\varepsilon = a$. Then phase field system (3.87), (3.88) can be written as

$$\frac{\partial \varphi}{\partial t} + \frac{\partial \theta}{\partial t} = \Delta \theta, \quad (x, t) \in Q, \tag{3.89}$$

$$\varepsilon^2 \frac{\partial \varphi}{\partial t} = \varepsilon^2 \Delta \varphi + \varphi - \varphi^3 + \varepsilon \kappa \theta. \tag{3.90}$$

Equations (3.89) and (3.90) are supplemented with the following initial and boundary conditions:

$$\varphi\big|_{t=0} = \varphi^0(x, \varepsilon), \quad \theta\big|_{t=0} = \theta^0(x, \varepsilon), \quad \varphi\big|_\Sigma = 1, \quad \theta\big|_\Sigma = \theta_b.$$

Here κ is a constant, $Q = (0, T) \times \Omega$, where $\Omega \subset \mathbb{R}^n, n \leq 3$, is a bounded domain with smooth (C^∞) boundary, $\Sigma = [0, T] \times \partial\Omega$, φ^0 and θ^0 are sufficiently smooth functions for $\varepsilon \geq \text{const} > 0$, and θ_b is a sufficiently smooth function. Let Γ_0 be a smooth surface of codimension 1, and let the intersection be $\Gamma_0 \cap \partial\Omega = \varnothing$. It is clear that Γ_0 divides Ω into two parts (domains) $\Omega_0^{\pm}$ so that $\Omega = \Omega_0^+ \cup \Gamma_0 \cup \Omega_0^-$. Let φ^0 and θ^0 be some special initial data such that $\varphi^0 = \pm 1 + O(\varepsilon)$ outside the ε-neighborhood of Γ_0 and $\theta^0 \in C(\Omega)$ (for details, see [6, 20, 48, 54] and below).

In this case, as $\varepsilon \to 0$, we obtain the modified Stefan problem (this problem is also called the Stefan–Gibbs–Thomson problem or the Stefan problem with surface tension):

$$\frac{\partial \theta^{\pm}}{\partial t} = \Delta \theta^{\pm}, \quad x \in \Omega_t^{\pm}, \quad t > 0, \tag{3.91}$$

$$\theta^{\pm}\big|_{t=0} = \theta_{\pm}^0(x), \quad x \in \Omega_0^{\pm}; \quad \theta^+\big|_{\Sigma} = \theta_b,$$

$$[\theta^{\pm}]\big|_{\Gamma_t} = 0, \quad \left[\frac{\partial \theta^{\pm}}{\partial \mathbf{n}}\right]\bigg|_{\Gamma_t} = -2v_{\mathbf{n}}, \tag{3.92}$$

$$\kappa_1 \theta^{\pm}\big|_{\Gamma_t} = \mathcal{K}_t - v_{\mathbf{n}}, \tag{3.93}$$

where $\theta_{\pm}^0(x) = \overline{\theta^0}(x)$ for $x \in \Omega_0^{\pm}$, $[f]\big|_{\Gamma_t}$, as was already defined above, is the jump of the function f along the free boundary Γ_t, $\mathbf{n}$ is the normal to Γ_t (outward normal with respect to Ω_t^-), $v_{\mathbf{n}}$ is the normal velocity of Γ_t, $\mathcal{K}_t = -\operatorname{div}(\mathbf{n})\big|_{\Gamma_t}$ is the mean curvature of the surface Γ_t, and $\kappa_1 = 3\kappa/\sqrt{2}$. We assume that $\Gamma_t \cap \partial\Omega = \varnothing$ for all $t \geq 0$.

Considering the situation with the problem for the mushy region, we meet the following two problems: there still does not exist a conventional model describing the mushy region, and moreover, it is unknown in advance which initial data regularized (with respect to the parameter ε) for the phase field model correspond to the situation with mushy region. We introduce the following basic assertions (hypotheses).

Assumption 3.8 The weak limit as $\varepsilon \to 0$ of the order function $\varphi(x, t, \varepsilon)$ is identically equal to zero for $x \in \Omega_t^*$ and corresponds to the mushy region Ω_t^*.

Assumption 3.9 In the domain $\Omega_{t,\varepsilon}^*$ corresponding to the "regularization" of the mushy region, the solution of the phase field system can be described in terms of the so-called "wave train" structure, i.e., a set of a large number of "small"-volume domains filled in succession with a liquid phase and a solid phase (or conversely).

It follows from Assumption 3.8 that, for almost all t, the limit order function $\overline{\varphi}$ belongs to $BV(\Omega)$ but $\overline{\varphi} \notin BVC(\Omega)$. Assumption 3.9 is based on the concept proposed in [38], where the "wave train" structure is used in the framework of the classical Stefan problem to describe the temperature in the mushy region. Such a structure is called a situation with a diffuse interface.

Thus, in the context of the above-cited hypotheses, taking the phase field model (3.89), (3.90) with surface tension as the initial model, we solve the following nonstandard problems. First, we solve the limiting problem describing the mushy region. Second, we seek initial data for Eqs. (3.89) and (3.90) for which the mushy region exists during a certain ε-independent time period.

The most important point is the procedure of passing to the limit as $\varepsilon \to 0$. It is clear that this transition can be performed only in the weak sense. Nevertheless, we must have a well-posed weak definition of the solution of problem (3.89), (3.90) (or the statement in the weak sense of the problem for the phase field system). The statement in the weak sense must satisfy the following natural conditions.

Assumption 3.10 In the situation with a fast varying localized perturbation (in particular, for φ^0 such that $\overline{\varphi^0} \in BVC(\Omega)$), such a definition must result in a well-posed limiting problem as $\varepsilon \to 0$.

Assumption 3.11 The statement in the weak sense must remain stable under small perturbations of the initial data.

The simplest method for verifying the well-posedness of a statement is widely used in the theory of distributions. Namely, the definition of weak solution (for example, the definition of the derivative in $\mathcal{D}'$, see [29]) for "good" functions (of class C^∞) must give the same results as the definition for the classical solution.

We perform such a verification for the phase field system in standard weak form obtained by almost exact repetition of the $\mathcal{D}'$-procedure for linear equations (also see [48]). But this showed that the problem in the standard statement is incorrect even in the situation with a fast varying localized perturbation. To avoid this problem, we propose a new formulation of the definition of weak solution of the phase field system (in fact, what is new here is only the weak form of the Allen–Cahn equation (3.90)). Here the starting point is the orthogonality condition or, (which is the same) solvability condition to the equation for the second term of asymptotic expansion φ^{as}, see Lemmas 3.2 and 3.3, which appears in the construction of the asymptotic solution in Sect. 3.3. This condition plays the central role in asymptotic analysis, and, in particular, only this condition leads to the Gibbs–Thomson condition on the free boundary.

The meaning of our definition of weak solution of the system of Eqs. (3.89), (3.90) is the integral approximation of this orthogonality condition. In this case, the conditions on the free boundary are weakly approximated by integral identities and we can pass to the limit as $\varepsilon \to 0$.

In what follows (see Sect. 3.4.1), we perform a test verification which shows that, in the situation with a fast varying localized perturbation on the interface between phases, the new statement permits passing to the limit as $\varepsilon \to 0$ and is stable. Moreover, the limit in the new statement of the problem leads to the weak statement of of the Gibbs–Thomson condition as a part of the definition of weak solution of the limiting problem. This implies that the new definition can be used in the situation with diffuse interface (see Sect. 3.4.2).

We note that these facts related to the weak statement of the phase field system are very important. In particular, to justify the stability of the new weak statement of the problem is a more important result than to solve the limiting problem in the situation with diffuse interface (mushy region). The point is that our approach to the definition of weak solution can be used to solve other similar problems (for example, the Musket problem, combustion waves, etc.). Namely, we use the orthogonality condition to construct a well-posed definition of weak solution of the regularized problem and, in the limit as $\varepsilon \to 0$, to construct the corresponding definition of weak solution of the limiting problem.

Let us make two remarks. Roughly speaking, the situation where $\overline{\varphi}$ is equal to zero on a set of nonzero measure is too "bad", because the zero value of the order function corresponds to an unstable solution of Allen–Cahn equation (3.90). Nevertheless, it is known in advance that such a solution can exist only in exceptional cases. In this situation, it is absolutely natural to impose very rigid conditions on the geometry of the domains Ω and Ω_t^* and on the initial and boundary conditions. We note that, since the temperature enters Eq. (3.90) for the order function with the coefficient ε, this equation is consistent with the Allen–Cahn equation with an accuracy of $O(\varepsilon)$. Correspondingly, the stable ($\overline{\varphi} = \pm 1$) and unstable ($\overline{\varphi} = 0$) solutions of this equation are independent of the temperature with an accuracy of $O(\varepsilon)$.

Another situation where the steady-state states of the order function "strongly" depend on the temperature is considered in [27, 28] (in this case, the coefficient of the temperature in the equation for the order function (3.90) is of the order $O(1)$). In this model, even in the simplest one-phase situation, the limit (as $\varepsilon \to 0$) solution cannot be described by the limit Stefan–Gibbs–Thomson problem, and the scenario of development of the initial perturbation of the unstable state is close to the scenario of development described by the van der Waals equation. Nevertheless, it was shown in [27, 28] that a structure of "wave train" type can arise from oscillating perturbations of an unstable stabilized state similar to that considered in this section.

From the standpoint of the theory of distributions, the problems with free boundary are problems about the motion of discontinuous solutions (singularities). Indeed, in the situation with fast varying localized perturbations, the limit order function is a Heaviside-type function ($\overline{\varphi} = 1$ on Ω_t^+ and $\overline{\varphi} = -1$ on Ω_t^-). The limit of the temperature remains continuous but has a weak discontinuity on the free boundary $\Gamma_t = \overline{\Omega_t^+} \cap \overline{\Omega_t^-}$.

In the theory of shock waves, the conditions arising on the shock wave front (necessary conditions for the solvability in the $\mathcal{D}'$ sense) are called Rankine–Hugoniot-type conditions. From this viewpoint, we can interpret conditions (3.92) as Rankine–Hugoniot-type conditions corresponding to the dynamics of the weak discontinuity of the limit temperature $\overline{\theta}$. If we use the solvability conditions in the sense of $\mathcal{D}'$ for the heat equation in (3.89) together with the new weak form of Eq. (3.90), then we obtain the limiting problem as an analog of the Rankine–Hugoniot condition for the system of phase field equations.

Since there does not exist a commonly accepted terminology, we refer to the conditions that appear in the interpretation of the problem with "wave train" for the mushy region as to Rankine–Hugoniot-type conditions.

3.4.1 *Weak Solutions and Rankine–Hugoniot-Type Conditions*

Starting from the classical definition of weak solution of linear differential equations (for example, see [55]), we define a weak solution of phase field system (3.89), (3.90).

Definition 3.1 Functions

$$\theta \in L^2\big(0, T; W_2^1(\Omega)\big) \cap L^\infty\big(0, T; L^2(\Omega)\big)$$

and

$$\varphi \in L^{\infty}\left(0, T; W_2^1(\Omega) \cap L^4(\Omega)\right)$$

are called weak solution of problem (3.89), (3.90) if, for any functions

$$\xi, g \in C^1(\overline{Q}), \quad \xi\big|_{\Sigma} = g\big|_{\Sigma} = 0, \quad \xi\big|_{t=T} = g\big|_{t=T} = 0, \tag{3.94}$$

the functions θ and φ satisfy the integral relations

$$I_{\theta} = \int_{Q} \left(\langle\nabla\theta, \nabla\xi\rangle - (\theta + \varphi)\xi_t\right) dx\,dt - \int_{\Omega} (\theta^0 + \varphi^0)\xi(x, 0)\,dx = 0, \tag{3.95}$$

$$I_{\varphi} = \int_{Q} \left(\varepsilon\langle\nabla\varphi, \nabla g\rangle - (\varphi - \varphi^3)g/\varepsilon - \kappa\theta g - \varepsilon\varphi g_t\right) dx\,dt$$

$$- \varepsilon \int_{\Omega} \varphi^0 g(x, 0)\,dx = 0. \tag{3.96}$$

Here by $\langle\cdot, \cdot\rangle$ we denote the inner product in $\mathbb{R}^n$, and by W_2^1, the Sobolev space.

As usual, relations (3.95), (3.96) for the functions θ and φ can be obtained by multiplying Eqs. (3.89) and (3.90) by test functions ξ and g and integrating by parts. A similar definition was also proposed in [48]. (The version of the phase field system without φ_t in Eq. (3.90) was considered in [48], but this is insignificant for the further analysis. In fact, the definition considered above was not used in [48].)

The above-introduced definition seems to be quite reasonable. Nevertheless, let us verify whether Definition 3.1 allows one to obtain the limiting problem (3.91)–(3.93). From this standpoint, we can use the fact that, in the situation with fast varying localized perturbation at $v_n \neq 0$, the solution obtained by asymptotic methods (see [20, 51], and (3.67), (3.68)) has the simple form

$$\theta_0^{as} = \frac{1}{2}\left(\tilde{\theta}^+(x, t) + \tilde{\theta}^-(x, t)\right) + \frac{1}{2}\left(\tilde{\theta}^+(x, t) - \tilde{\theta}^-(x, t)\right)\chi(\eta), \tag{3.97}$$

$$\varphi_1^{as} = \chi(\eta) + \varepsilon\left\{\frac{\kappa}{2}\theta_0^{as} + \omega(\eta, x)\right\}, \tag{3.98}$$

where $\eta = s(x, t, \varepsilon)/\varepsilon$ and $\chi(\eta) = \tanh\left(\eta/\sqrt{2}\right)$. Here θ_0^{as} is the leading term of the asymptotic expansion of θ^{as} for the temperature, φ_1^{as} denotes the first two terms of the asymptotic expansion of φ^{as} for the order function, where θ^{as} and φ^{as} obtained in Sect. 3.3. The function of the distance s between the free boundary Γ_t and points $x \in \Omega$ $s(x)$ becomes

$$s(x, t, \varepsilon) = s_0(x, t) + \varepsilon s_1(x).$$

The function s_0 has the form $s_0 = (t + \psi(x))/|\nabla\psi|$. Here $\psi(x)$ and the auxiliary functions $\theta^{\pm}$ form the solution of problem (3.91)–(3.93) with

$$\Gamma_t = \{x, \psi(x) = -t\}, \quad \mathbf{n} = \nabla\psi/|\nabla\psi|, \tag{3.99}$$

$$v_{\mathbf{n}} = -1/|\nabla\psi|, \quad \mathcal{K}_t = -\operatorname{div}\mathbf{n}.$$

We also denote the continuous extensions across $\Gamma(t)$ of the functions $\theta^{\pm}$ by $\theta_c^{\pm}$, $\omega = \omega(\eta, x) \in \mathbf{S}, \mathbf{S} = C^{\infty}(\Omega; \mathcal{S}(\mathbb{R}^1_{\eta}))$, $\mathcal{S}$ is the Schwartz space, and $s_1 = \psi_1(x)/|\nabla\psi|$ is a smooth function. It was shown in [20] and Sect. 3.3 that the asymptotic solution in the form (3.97), (3.98) is independent of the extension method, construction of the function ψ_1, and the higher-order terms in the asymptotic expansion.

Moreover, in [10, 21] with regard to some conditions [50] ensuring the classical solvability of problem (3.91)–(3.93), it is also proved that if

$$\left\|\theta^0 - \theta^{as}\big|_{t=0}; L^2(\Omega)\right\| + \left\|\varphi^0 - \varphi^{as}\big|_{t=0}; L^2(\Omega)\right\| \leq c\varepsilon^{\mu}, \quad \mu \geq 3/2,$$

then for sufficiently small ε, there exists a unique solution of problem (3.89), (3.90) such that

$$\left\|\varphi - \varphi^{as}; C(0, T; L^2(\Omega))\right\| + \left\|\theta - \theta^{as}; L^2(Q)\right\| \leq c_1\varepsilon^{\mu} \tag{3.100}$$

with constants c and c_1 independent of ε.

Note that, for any function $\omega(\eta, x) \in \mathbf{S}$ and any $s \in C^2(\overline{Q})$ such that $|\nabla s|\big|_{s=0} \neq 0$,

$$\left\|\omega\big(s(x, t, \varepsilon)/\varepsilon, x\big); C(0, T; L^2(\Omega))\right\| \leq \operatorname{const} \varepsilon^{1/2}. \tag{3.101}$$

In particular, the estimates (3.99)–(3.101) mean that the solution is stable with respect to small (of the order of $O(\varepsilon)$ in $C(\overline{Q})$ and of the order of $O(\varepsilon^{3/2})$ in $C(0, T; L^2(\Omega))$) fast decreasing localized perturbations of initial data. In particular, with these perturbations taken into account, the solution has the form

$$\theta = \theta_0^{as} + \varepsilon\theta_{\omega_1}, \tag{3.102}$$

$$\varphi = \chi(\eta) + \varepsilon\left\{\frac{\kappa}{4}\theta_0^{as} + \varepsilon^{1/2}\omega_1(\eta, x, t, \varepsilon)\right\}, \tag{3.103}$$

where θ_{ω_1} and ω_1 are corrections such that

$$\left\|\theta_{\omega_1}; C(0, T; L^2(\Omega))\right\| \leq \operatorname{const},$$

$$\left\|\varepsilon^{1/2}\omega_1(\eta, x, t, \varepsilon); C(\overline{Q})\right\| \leq \operatorname{const},$$

$$\left\|\omega_1(\eta, x, t, \varepsilon); C(0, T; L^2(\Omega))\right\| \leq \operatorname{const},$$

and the limiting problem is still the modified Stefan–Gibbs–Thomson problem. But this fact does not agree with Definition 3.1. In other words, we show that Definition 3.1 is unstable with respect to small perturbations of the initial data.

We consider problem (3.89), (3.90) with the initial data

$$\varphi\Big|_{t=0} = \chi(\eta^0) + \varepsilon\left\{\frac{\kappa}{4}\theta_0^{\mathrm{as}}\Big|_{t=0} + \omega_1^0(\eta^0, x)\right\}, \qquad (3.104)$$

$$\theta\Big|_{t=0} = \theta_0^{\mathrm{as}}\Big|_{t=0}, \qquad (3.105)$$

where $\eta^0 = s^0/\varepsilon$, $s^0(x)$ is the function of the distance from Γ_0, $\omega_1^0(\eta^0, x) \in \mathbf{S}$ differ from the fixed function $\omega(\eta, x)\big|_{\eta=\eta^0}$ in the formula for the asymptotic solution φ_1^{as} with an accuracy of $O(\varepsilon^2)$, see (3.98).

First, we note that the weak limit of the function

$$\varepsilon^{-1}\omega\left(\frac{t+\psi}{\varepsilon}, x\right),$$

where $\omega(\eta, x) \in \mathbf{S}$, is the Dirac δ function on the surface

$$\mathcal{T} = \bigcup_{t\in[0,T]} \Gamma_t \subset Q.$$

Lemma 3.4 *We assume that* $\omega(\eta, x) \in \mathbf{S}$, *the function* $\psi(x) \in C^2(\overline{\Omega})$ *is such that* $|\nabla\psi| \neq 0$, *and* $\mathrm{dist}(\Gamma_t, \partial\Omega) \geq \mathrm{const} > 0$. *Then, for any function* $g \in C^1(\overline{Q})$,

$$\lim_{\varepsilon\to 0} \varepsilon^{-1}\int_Q \omega\left(\frac{s}{\varepsilon}, x\right)g(x, t)\,\mathrm{d}x\mathrm{d}t = \int_{\Omega_T} A_\omega(x)\check{\beta}^{-1}(x)g(x, -\psi)\,\mathrm{d}x, \qquad (3.106)$$

where $s = (t + \psi)\check{\beta} + \varepsilon s_1$, $\check{\beta} = |\nabla\psi|^{-1}$, Ω_T *is the domain between the surfaces* Γ_0 *and* Γ_T, *and*

$$A_\omega - \int_{-\infty}^{\infty} \omega(\eta, x)\,\mathrm{d}\eta.$$

The proof follows from commonly known facts of the generalized function theory [29].

Obviously, the right-hand side of (3.106) can be rewritten as

$$\int_{\Omega_T} A_\omega(x)\check{\beta}^{-1}(x)g(x, -\psi)\,\mathrm{d}x = \left(A_\omega\delta(\mathcal{T}), g\right) \overset{\mathrm{def}}{=} \int_{\mathcal{T}} A_\omega g\,\mathrm{d}\sigma,$$

where $\delta(\mathcal{T})$ is the δ function on the surface $\mathcal{T}$ and $\check{\beta}^{-1}dx$ is the Leray measure $d\sigma$ on $\mathcal{T}$ defined by the relation

$$\mathrm{d}\big(\check{\beta}(t + \psi)\big) \wedge d\sigma\big|_{t=-\psi} = \mathrm{d}t\mathrm{d}x,$$

also see [29].

At first, let us calculate the integrals in formulas (3.95), (3.96). From (3.97), (3.98), (3.102), (3.103), (3.106) as $\varepsilon \to 0$ we obtain

$$\int_Q \varphi \xi_t \, \mathrm{d}x \mathrm{d}t + \int_\Omega \varphi^0 \xi(x,0) \, \mathrm{d}x$$

$$= \varepsilon^{-1} \int_Q (\check{\beta}\dot{\chi} + \varepsilon v)\xi \, \mathrm{d}x\mathrm{d}t + O(\varepsilon) \to 2 \int_{\mathcal{T}} \check{\beta}\xi \, \mathrm{d}\sigma,$$

where $v = v_0(s/\varepsilon, x) + O(\varepsilon)$ and the function $v_0(\eta, x) \in \mathbf{S}$ can be calculated by using (3.102), (3.103) with regard to the relation $A_{\dot{\chi}} = 2$. Thus, as $\varepsilon \to 0$, we have

$$I_\theta \to \int_0^T \left\{ - \int_{\Omega_t^+} (\Delta\theta^+ - \theta_t^+)\xi \, \mathrm{d}x - \int_{\Omega_t^-} (\Delta\theta^- - \theta_t^-)\xi \, \mathrm{d}x \right\} \mathrm{d}t$$

$$- \int_{\mathcal{T}} \left(\frac{\partial\theta^+}{\partial\mathbf{n}} - \frac{\partial\theta^-}{\partial\mathbf{n}} - 2\check{\beta} \right)\xi \, \mathrm{d}\sigma = 0.$$

This leads to Eqs. (3.89) and (3.90) and the second Stefan condition (3.92). Moreover, substituting formulas (3.102), (3.103) into (3.96), we rewrite (3.96) as

$$I_\varphi = - \varepsilon^{-1} \int_Q (\ddot{\chi} + \chi - \chi^3)g \, \mathrm{d}x\mathrm{d}t$$

$$- \int_Q (F + \varepsilon^{1/2}(1 - 3\chi^2)\omega_1)g \, \mathrm{d}x\mathrm{d}t$$

$$+ \varepsilon^{1/2} \int_Q \langle \varepsilon\nabla\omega_1, \varepsilon\nabla g \rangle \, \mathrm{d}x\mathrm{d}t + O(\varepsilon) = 0, \tag{3.107}$$

where

$$F = 2\Big(\eta\langle\nabla\psi, \nabla\check{\beta}\rangle + \check{\beta}^2\langle\nabla\psi, \nabla\psi_1\rangle \Big)\ddot{\chi}$$

$$+ \Big(2\langle\nabla\psi, \nabla\check{\beta}\rangle + \check{\beta}(\Delta\psi - 1) \Big)\dot{\chi} + \frac{3}{2}\kappa\theta_0^{\mathrm{as}}(1 - \chi^2). \tag{3.108}$$

The first integral in formula (3.107) is equal to zero, because

$$\chi = \tanh\left(\frac{\eta}{\sqrt{2}} \right).$$

Calculating the remaining integrals, we obtain

$$I_\varphi = -\varepsilon \int\limits_{\Omega_T} \left\{ 2\check{\beta}(\Delta\psi - 1) + \frac{3}{2}\kappa\theta^\pm(x, -\psi)A_{(1-\chi^2)} \right\} g(x, -\psi)\check{\beta}^{-1}\mathrm{d}x$$

$$- \varepsilon^{1/2} \int\limits_Q f_1(\eta, x, t, \varepsilon)g\,\mathrm{d}x\mathrm{d}t + O(\varepsilon^{3/2}) = 0, \tag{3.109}$$

where $f_1(\eta, x, t, \varepsilon) = \left(1 - 3\chi^2(\eta)\right)\omega_1(\eta, x, t, \varepsilon)$ and

$$A_{(1-\chi^2)} = \int\limits_{-\infty}^{\infty} (1 - \chi^2)\,\mathrm{d}\eta = 2\sqrt{2}.$$

It follows from the preceding that

$$\left\| f_1; C\left(0, T; L^2(\Omega)\right) \right\| \le \text{const}$$

uniformly in ε.

Nevertheless, since the Allen–Cahn operator linearized on φ^{as} has a continuous spectrum, one cannot guarantee that the function $f_1(x, t, \varepsilon)$ will be localized near the free boundary with an accuracy of $O(\sqrt{\varepsilon})$ as $\varepsilon \to 0$.

If the function $f_1\left((t + \psi)/\varepsilon, x, t, \varepsilon\right)$ mod $O(\sqrt{\varepsilon})$ is not localized, then despite (3.93), it follows from (3.109) that

$$\lim_{\varepsilon \to 0} f_1\left(\frac{t + \psi}{\varepsilon}, x, t, \varepsilon\right) = 0, \tag{3.110}$$

where the limit is understood in the sense of $\mathcal{D}'(Q)$ and, in general, this condition (3.110) is satisfied in the whole domain Q.

If the function $f_1\left((t + \psi)/\varepsilon, x, t, \varepsilon\right)$ mod $O(\sqrt{\varepsilon})$ is localized, then it is localized with an accuracy of $O(\varepsilon^{1/2})$ in $\mathcal{D}'(Q)$. In this case, with regard to (3.109), the sum of integrands tends to zero and, taking (3.93) and (3.99) into account, we obtain

$$\kappa_1\theta^\pm(x, -\psi) = v_{\mathbf{n}}(\Delta\psi - 1) - \frac{1}{2}\overline{A}_{f_1}\Big|_{t=-\psi}, \tag{3.111}$$

where $\overline{A}_{f_1}$ has the form

$$\overline{A}_{f_1} = \lim_{\varepsilon \to 0} \int\limits_{-\infty}^{\infty} f_1(\eta, x, t, \varepsilon)\,\mathrm{d}\eta.$$

Obviously, both conditions (3.110) and (3.111) differ from Gibbs–Thomson condition (3.93). Moreover, we here have the unknown function $\overline{A}_{f_1}$ (or $\bar{f}_1$) depending on the first-order correction ω_1 in (3.102), (3.103). Therefore (in contrast to (3.93)), condition (3.111) (or (3.110)) does not complete system (3.89), (3.90) to a closed system of equations. Thus, we cannot calculate the position of the free boundary. We can consider the structure of the fast varying localized solution up to $O(\varepsilon)$, i.e., by changing the initial conditions as

$$\theta\big|_{t=0} = \theta_1^{as}\big|_{t=0},$$
$$\varphi\big|_{t=0} = \varphi_2^{as}\big|_{t=0} + \varepsilon^2 \omega_2^0(\eta^0, x),$$

where $\omega_2^0 \in \mathbf{S}$ is an arbitrary function, θ_1^{as} and φ_2^{as} are the right hand sides of (3.97), (3.98) plus their first corrections. Then we see that $\varepsilon^{1/2}\omega_1 = \omega$ is the function in representations (3.97), (3.98). Thus, the last two integrals in formula (3.107) tend to zero by the definition of ω in the construction of the asymptotic solution. Under the assumption that the terms of the order $O(\varepsilon^2)$ are equal to zero, we obtain linearization (3.111) for the localized f_2. But this relation contains a new unknown function $\overline{A}_{f_2}$. Continuing this procedure, we obtain an infinite system of connected relations of the form similar to the Rankine–Hugoniot conditions for shock waves. We have already noted that we cannot calculate the dynamics of the free boundary Γ_t from this system of relations. Moreover, we cannot disjoin this system using the exact solution of the limiting problem in $\Omega_t^{\pm}$.

But we can change the definition of weak solution of the phase field system so that a necessary condition for the existence of a solution with fast varying localized perturbation follows precisely from Gibbs–Thomson condition (3.93) instead of condition (3.110) (or condition (3.111)).

Definition 3.2 A pair of functions

$$\theta \in L^2\big(0, T; W_2^1(\Omega)\big) \cap L^\infty\big(0, T; L^2(\Omega)\big),$$
$$\varphi \in W_2^{2,1}(Q) \cap L^\infty\big(0, T; W_2^1(\Omega) \cap L^4(\Omega)\big)$$

is called a weak solution of problem (3.89), (3.90) if, for any functions $\xi(x, t)$ and $g(x, t) = \big(g_1(x, t), \dots, g_n(x, t)\big)$ satisfying condition (3.94), the functions θ and φ satisfy (3.95) and the integral identity

$$J_\varphi = \varepsilon \int_Q \varphi_t \langle g, \nabla\varphi \rangle \, dxdt - \int_Q e_\varepsilon(\varphi) \, \mathrm{div}\, g \, dxdt$$
$$+ \int_Q \big(\varepsilon \langle \nabla\varphi, g_x \nabla\varphi \rangle + \kappa\varphi \, \mathrm{div}\, (g\theta)\big) \, dxdt = 0, \tag{3.112}$$

where

$$e_\varepsilon(\varphi) = \frac{\varepsilon|\nabla\varphi|^2}{2} + \frac{W(\varphi)}{\varepsilon}, \quad W(\varphi) = \frac{(\varphi^2 - 1)^2}{4},$$

and g_x is the matrix with entries $(g_x)_{ik} = \partial g_i / \partial x_k$.

Identity (3.112) can be obtained for smooth functions φ and θ by multiplying Eqs. (3.90) by $\langle g, \nabla\varphi\rangle$ and integrating by parts. We note that the idea of this definition was used in [48].

First, we verify Definition 3.2 just as we verified Definition 3.1. It is clear that it suffices to calculate the integral J_φ as $\varepsilon \to 0$. Using representations (3.102), (3.103), we obtain

$$
\begin{aligned}
J_\varphi = {}& \varepsilon^{-2} \int_Q (\ddot{\chi} + \chi - \chi^3)(\check{\beta}\langle g, \nabla\psi\rangle\dot{\chi} + \varepsilon G)\,\mathrm{d}x\mathrm{d}t \\
& + \varepsilon^{-1} \int_Q \check{\beta}\langle g, \nabla\psi\rangle(\varepsilon^{1/2}\omega_1\hat{L} + F)\dot{\chi}\,\mathrm{d}x\mathrm{d}t + O(\varepsilon^{1/2}) = 0, \quad (3.113)
\end{aligned}
$$

where

$$
\hat{L} = \frac{\partial^2}{\partial\eta^2} + 1 - 3\chi^2,
$$

expression (3.108) is also used, and the function G is bounded in $C\big(0, T; L^2(\Omega)\big)$. Because $\chi = \tanh(\eta/\sqrt{2})$ and $L\dot{\chi} = 0$, it follows from Lemma 3.4 that relation (3.113) can be transformed as

$$
J_\varphi = \int_{\Omega_T} \left\{ \int_{-\infty}^{\infty} F\dot{\chi}\,\mathrm{d}\eta \right\} \langle g, \nabla\psi\rangle \Big|_{t=-\psi}\,\mathrm{d}x + O(\varepsilon^{1/2}) = 0. \quad (3.114)
$$

It is clear that to satisfy (3.114), it is necessary that the expression in curly brackets tend to zero. Using explicit form (3.108) for the function F, we here obtain just the Gibbs–Thomson condition (3.93).

Expanding ψ^{as}, θ^{as}, and J_φ in series in ε, we can easily see that the obtained system of equations is triangular in the sense that, for each N, the first N equations of the system contain N unknown functions. Thus, starting from Definition 3.1, we see that the contribution of small fast decreasing localized perturbations to the solution of the equation of motion of the free boundary is of the order $O(1)$. If we use Definition 3.2, then the lower-order terms do not make contributions to this law of motion. In other words, the limiting problem understood in the sense of Definition 3.2 is stable under small perturbations when one passes to the limit as $\varepsilon \to 0$.

But we have verified Definition 3.2 only for special initial data (3.104), (3.105) with fast decreasing localized perturbations. Let us consider more general initial data.

Let $\overline{\varphi^0}(x) \in BVC(\Omega)$. Following [41] (also see [45]), we assume that $\varphi^0(x, \varepsilon) \in W_2^1(\Omega)$ is a family of functions such that, as $\varepsilon \to 0$,

$$
\varphi^0(x, \varepsilon) \to \overline{\varphi^0}(x) \text{ in } L^2(\Omega), \quad E_\varepsilon\big(\varphi^0(x, \varepsilon)\big) \to \frac{\sqrt{2}}{3} \int_\Omega \mathrm{d}|\nabla\overline{\varphi^0}|, \quad (3.115)
$$

where

$$E_\varepsilon(\varphi) = \int\limits_\Omega e_\varepsilon(\varphi)\,\mathrm{d}x.$$

Then using the technique developed in [41] and repeating the construction in [48], we can prove the following assertion (also see [54]).

Theorem 3.3 *For the initial data*

$$\theta^0 = \overline{\theta^0} \in L^2(\Omega) \quad and \quad \varphi^0(x,\varepsilon) \in W_2^1(\Omega)$$

satisfying conditions (3.115), there exists a weak solution of problem (3.89), (3.90) (in the sense of Definition 3.2) such that

$$\varphi \in W_2^{2,1}(Q) \cap L^\infty\big(0, T; W_2^1(\Omega) \cap L^4(\Omega)\big),$$
$$\theta \in L^\infty\big(0, T; L^2(\Omega)\big) \cap L^2\big(0, T; W_2^1(\Omega)\big).$$

Moreover, if for all $v \in BVC(\Omega)$, the functions $\overline{\varphi^0}$ and $\overline{\theta^0}$ satisfy the variational problem

$$\frac{\sqrt{2}}{3} \int\limits_\Omega \mathrm{d}|\nabla\overline{\varphi^0}|(x) - (\overline{\theta^0}, \overline{\varphi^0}) \le \frac{\sqrt{2}}{3} \int\limits_\Omega \mathrm{d}|\nabla v|(x) - (\overline{\theta^0}, v),$$

then one can pass to the limit as $\varepsilon \to 0$ in integral identities (3.95) and (3.112):

$$\lim_{\varepsilon\to0} I_\theta = \int\limits_Q \big(\langle\nabla\overline{\theta}, \nabla\xi\rangle - (\overline{\theta} + \overline{\varphi}), \xi_t\big)\,\mathrm{d}x\mathrm{d}t + \int\limits_\Omega (\overline{\theta^0} + \overline{\varphi^0})\xi(x,0)\,\mathrm{d}x = 0,$$

$$(3.116)$$

$$\lim_{\varepsilon\to0} J_\varphi = \int\limits_0^T \mathrm{d}t \int\limits_\Omega v_{\mathbf{n}}\langle\mathbf{n}, g\rangle\,|\nabla_x\overline{\varphi}|(t, \mathrm{d}x)$$

$$- \int\limits_0^T \mathrm{d}t \int\limits_\Omega \big(\mathrm{div}\,g - \langle\mathbf{n}, g_x\mathbf{n}\rangle\big)\,|\nabla_x\overline{\varphi}|(t, \mathrm{d}x)$$

$$+ \kappa \int\limits_Q \overline{\varphi}\,\mathrm{div}\,(g\overline{\theta})\,\mathrm{d}x\mathrm{d}t = 0.$$

$$(3.117)$$

Here the limit functions are

$$\overline{\theta} \in L^\infty(0, T; L^2(\Omega)) \cap L^2(0, T; W_2^1(\Omega)),$$
$$\overline{\varphi} \in L^\infty\big(0, T; BVC(\Omega)\big),$$

and the generalized normal velocity $v_\mathbf{n}$ and the unit vector $\mathbf{n}$ are defined by the Radon–Nikodym theorem as

$$\mathrm{d}\overline{\varphi} = (-v_\mathbf{n}, \mathbf{n})^\mathrm{T}\,\mathrm{d}t\,|\nabla_x\overline{\varphi}|(t, \mathrm{d}x),$$

where $\mathrm{d}\overline{\varphi}$ is the vector measure of the generalized gradient

$$\left(\frac{\partial}{\partial t}\overline{\varphi}, \nabla_x\overline{\varphi}\right)^\mathrm{T},$$

$v_\mathbf{n} \in L^\infty\big(Q, \mathrm{d}t\,|\nabla_x\overline{\varphi}|(t, \mathrm{d}x)\big)$, and $(a, b)^\mathrm{T}$ is the vector with components a and b.

Thus, using Definition 3.2 in the situation with fast varying localized perturbation on the interface between phases, we obtain the desired limiting problem (also see [40, 48, 54]). The result obtained in this section shows that Definition 3.2 is well posed.

3.4.2 Solutions of "Wave Train" Type and the Corresponding Limiting Problem

We consider a more general problem, where $\overline{\varphi^0} \in BV$ but $\overline{\varphi^0} \notin BVC$. This it the so-called on-boundary situation related to the well-known problem with "mushy region". Our consideration is based on the concept introduced in [17, 38] (also see the references therein), where the "mushy region" is represented as a set of a large number M of "solid" and "liquid" domains of small volume v_ε ($M = M(\varepsilon) \to \infty$ and $v_\varepsilon \to 0$ as $\varepsilon \to 0$); we present a macroscopic description by calculating the weak limit as $\varepsilon \to 0$.

First, we describe the original geometric structure. We assume that, for $t = 0$, the domain Ω contains only pure "solid" and pure "liquid" domains $\Omega_{0,\varepsilon}^\pm$ and the melting zone $\Omega_{0,\varepsilon}^*$ filled with "solid" and "liquid" domains $\Omega_{0,\varepsilon}^i$, $i = 1, 2, \ldots, M$, where M is an even number. To simplify the problem, we consider the case of quasispherical symmetry. Let $\Gamma_{0,\varepsilon}^i$, $i = 1, \ldots, M - 1$, be the interface between the domains $\Omega_{0,\varepsilon}^i$ such that $\partial\Omega_{0,\varepsilon}^i = \Gamma_{0,\varepsilon}^{i-1} \cup \Gamma_{0,\varepsilon}^i$, and let $\Gamma_{0,\varepsilon}^0 = \partial\Omega_{0,\varepsilon}^-$ and $\partial\Omega_{0,\varepsilon}^+ = \Gamma_{0,\varepsilon}^M \cup \partial\Omega$. By $D_{0,\varepsilon}^i$ we denote the domains bounded by $\Gamma_{0,\varepsilon}^i$ and assume that $D_{0,\varepsilon}^i \subset D_{0,\varepsilon}^{i+1}$, $i = 0, \ldots, M$, where $D_{0,\varepsilon}^0 = \Omega_{0,\varepsilon}^-$ and $D_{0,\varepsilon}^{M+1} \equiv \Omega$. We also assume that $\Gamma_{0,\varepsilon}^i$ are smooth surfaces of codimension 1 such that

$$c_1\varepsilon^\alpha \le \mathrm{dist}(\Gamma_{0,\varepsilon}^{k-1}, \Gamma_{0,\varepsilon}^k) \le c_2\varepsilon^\alpha, \tag{3.118}$$

$$c_1^\pm \le |\Omega_{0,\varepsilon}^\pm| \le c_2^\pm; \quad \mathrm{dist}(\Gamma_{0,\varepsilon}^M, \partial\Omega) \ge c_3,$$

where $k = 1, \ldots, M$; $\alpha \in (0, 1)$ and $c_j^\pm$, $c_j > 0$, $j = 1, 2$, are constants independent of ε. We also assume that the following geometric conditions are satisfied.

Assumption 3.12 The system of surfaces $\Gamma_{0,\varepsilon}^i$ such that $\Gamma_{0,\varepsilon}^i \in C^3$ uniformly in $\varepsilon \in [0, \varepsilon_0]$ for all $i = 0, 1, \ldots, M$, $M \to \infty$ and $M\varepsilon^\alpha \to L = \text{const}$ as $\varepsilon \to 0$, fills in the limit a certain domain Ω_0^* bounded by C^3-surfaces Γ_0^- and Γ_0^+.

If Assumption 3.12 is satisfied, then there exists a function $s^0(x, \varepsilon) \in C^3(\overline{\Omega})$ such that each $\Gamma_{0,\varepsilon}^i$ is a level surface of this function.

We choose some special initial data for the prelimit phase field system based on an implicit formula for the asymptotic solution in the case of fast varying localized perturbation. Obviously, (3.118) implies that the initial data can be locally represented in a form similar to (3.97), (3.98) for $t = 0$.

Moreover, formulas (3.97), (3.98) show that there is no interaction (up to $O(\varepsilon^\infty)$) between waves of the form tanh until the distance between neighborhoods of the interfaces becomes less than $O(\varepsilon^{1-\delta})$ for any constant $\delta > 0$. Nevertheless, for sufficiently small t, the asymptotic solution of the problem in question remains a superposition of local solutions (3.97), (3.98). Namely,

$$\varphi_1^{\text{as}}(x, t, \varepsilon) = \sum_{i=0}^{M} (-1)^i \chi(\eta_i) + \varepsilon \left\{ \frac{\kappa}{2} \hat{\theta}_0^{\text{as}}(x, t, h) + \sum_{i=0}^{M} \omega_i(\eta_i, x) \right\} \qquad (3.119)$$

(for the definitions of $\eta_i = \eta_i(x, t, \varepsilon)$, see below). Here, for convenience, we assume that there exist two functions $s^{(1)}(x, t, \varepsilon)$ and $s^{(2)}(x, t, \varepsilon)$ which, respectively, describe the surfaces $\Gamma_{t,\varepsilon}^i$ with odd and even numbers as $t \geq 0$. By $\Omega_{t,\varepsilon}^i$ we denote the domains between the surfaces $\Gamma_{t,\varepsilon}^{i-1}$ and $\Gamma_{t,\varepsilon}^i$, $i = 1, \ldots, M$, and write

$$\Omega_{t,\varepsilon}^* = \bigcup_{i=1}^{M} \Omega_{t,\varepsilon}^i.$$

Constructing a formal asymptotic solution, we define

$$s^{(j)}(x, t, \varepsilon) = s_0^{(j)}(x, t, h) + \varepsilon c_1^{(j)}(x, t, h),$$

where $h = \varepsilon^\alpha$, $\varepsilon \in [0, \varepsilon_0]$, $j = 1, 2$, so that $|\nabla_x s^{(j)}| > 0$ uniformly in $x \in \Omega_{t,\varepsilon}^*$ for all $h \in [0, h_0 = \varepsilon_0^\alpha]$ and

$$\Gamma_{t,\varepsilon}^i = \left\{ x, s_0^{(n_i)}(x, t, h) = ih \right\}, \qquad (3.120)$$

where $n_i = 1$ for $i = 2k$ and $n_i = 2$ for $i = 2k + 1$, $0 \leq i \leq M$. Obviously, $s^{(1)}\big|_{t=0} = s^{(2)}\big|_{t=0} = s^0(x, \varepsilon)$ and with an accuracy of $O(\varepsilon)$ the outward normals on $D_{t,\varepsilon}^i$ have the form

$$\mathbf{n}_i = \frac{\nabla s_0^{(n_i)}}{|\nabla s_0^{(n_i)}|}\bigg|_{\Gamma_{t,\varepsilon}^i}.$$

We use the following notation in (3.119):

$$\eta_i = \frac{s^{(n_i)}(x, t, \varepsilon) - ih}{|\nabla s_0^{(n_i)}|\varepsilon} \in \mathbf{S},$$

$\omega_i \in \mathbf{S}$, and $\hat{\theta}_0^{\mathrm{as}}$ is a smooth function (for a fixed $\varepsilon > 0$) whose local representation has the form

$$\hat{\theta}_0^{\mathrm{as}} = \frac{1}{2}(\theta_{i-1,c} + \theta_{i,c}) + \frac{1}{2}(\theta_{i-1,c} - \theta_{i,c})\chi(\eta_i), \quad x \in \Omega_{t,\varepsilon}^{i-1} \cup \Omega_{t,\varepsilon}^{i}.$$

As in the situation with a fast varying localized perturbation, here $\theta_{i,c}$ are sufficiently smooth continuations of auxiliary functions $\theta_i = \theta_i(x, t, h)$. In turn, the family of functions $\{\theta_i\}$ and $s_0^{(j)}$, $j = 1, 2$, is defined as the solution of the following set of Stefan–Gibbs–Thomson problems

$$\frac{\partial \theta_i}{\partial t} = \Delta\theta_i, \quad x \in \Omega_{t,\varepsilon}^{i}, \quad t > 0, \tag{3.121}$$

$$\theta_{i-1}\big|_{\Gamma_{t,\varepsilon}^{i-1}-0} = \theta_i\big|_{\Gamma_{t,\varepsilon}^{i-1}+0}, \quad \theta_i\big|_{\Gamma_{t,\varepsilon}^{i}-0} = \theta_{i+1}\big|_{\Gamma_{t,\varepsilon}^{i}+0}, \tag{3.122}$$

$$\frac{\partial \theta_{i-1}}{\partial \mathbf{n}_{i-1}}\bigg|_{\Gamma_{t,\varepsilon}^{i-1}-0} - \frac{\partial \theta_i}{\partial \mathbf{n}_{i-1}}\bigg|_{\Gamma_{t,\varepsilon}^{i-1}+0} = (-1)^{i+1}2v_{\mathbf{n}_{i-1}}, \tag{3.123}$$

$$\frac{\partial \theta_i}{\partial \mathbf{n}_i}\bigg|_{\Gamma_{t,\varepsilon}^{i}-0} - \frac{\partial \theta_{i+1}}{\partial \mathbf{n}_i}\bigg|_{\Gamma_{t,\varepsilon}^{i}+0} = (-1)^{i}2v_{\mathbf{n}_i}, \tag{3.124}$$

$$(-1)^{i-1}\kappa_1\theta_i\big|_{\Gamma_{t,\varepsilon}^{i-1}-0} = \mathcal{K}_t^{i-1} - v_{\mathbf{n}_{i-1}}, \tag{3.125}$$

$$(-1)^{i}\kappa_1\theta_i\big|_{\Gamma_{t,\varepsilon}^{i}+0} = \mathcal{K}_t^{i} - v_{\mathbf{n}_i}, \tag{3.126}$$

$$i = 0, 1, \dots, M + 1,$$

supplemented with the initial and boundary conditions (in $\partial\Omega \times [0, T]$). We put $\Gamma_{t,\varepsilon}^{-1} = \Gamma_{t,\varepsilon}^{M+1} = \varnothing$ so that the first condition in (3.122)–(3.126) disappears for $i = 0$ and the second condition in (3.122)–(3.126) disappears for $i = M + 1$. Moreover,

$$v_{\mathbf{n}_i} = -\left(\left|\nabla s_0^{(n_i)}\right|\right)^{-1}\frac{\partial s_0^{(n_i)}}{\partial t}\bigg|_{\Gamma_{t,\varepsilon}^{i}},$$

$$\mathcal{K}_t^{i} = -\operatorname{div}\mathbf{n}_i\big|_{\Gamma_{t,\varepsilon}^{i}},$$

and by $\Omega_{t,\varepsilon}^0 = \Omega_{t,\varepsilon}^-$ we denote the domains bounded by $\Gamma_{t,\varepsilon}^0$, and by $\Omega_{t,\varepsilon}^{M+1} = \Omega_{t,\varepsilon}^+$, the domain bounded by $\Gamma_{t,\varepsilon}^M$ and $\partial\Omega$. The small corrections $c_1^{(j)}(x, t, h)$ and the corrections of the order $O(\varepsilon)$ for the temperature can be obtained from the linearized Stefan–Gibbs–Thomson problem [20].

For a fixed $\varepsilon > 0$ and sufficiently small $t > 0$, the classical solvability of problem (3.121)–(3.126) can be proved in detail [21, 50]. At the same time, since the classical solvability of the Stefan–Gibbs–Thomson problem leads to consistency conditions on all initial surfaces $\Gamma^i_{0,\varepsilon}$, it is impossible for the limiting problems to formulate the conditions for the initial temperature $\theta^0(x, \varepsilon)$ so that these conditions be meaningful as $\varepsilon \to 0$. But we can avoid this problem if we find a model problem for the weak limit of the temperature as $\varepsilon \to 0$.

Thus, we choose the following initial data:

$$\varphi\big|_{t=0} = \varphi_1^{\text{as}}(x, 0, \varepsilon) + O(\varepsilon^2),$$
$$\theta\big|_{t=0} = \hat{\theta}_0^{\text{as}}(x, 0, \varepsilon) + O(\varepsilon), \tag{3.127}$$
$$s^{(j)}\big|_{t=0} = s^0(x, \varepsilon),$$

where $\hat{\theta}_0^{\text{as}}(x, 0, \varepsilon)$, $\varphi_1^{\text{as}}(x, 0, \varepsilon)$, and $s^0(x, \varepsilon)$ are smooth functions such that the consistency conditions are satisfied for a fixed $\varepsilon > 0$. But we can sharpen these conditions by calculating the limiting problem (see below).

Just as for the evolution of the solution, we have two significantly distinct situations (determined by the initial data):

$$\frac{\partial s_0^{(1)}}{\partial t} \frac{\partial s_0^{(2)}}{\partial t} < 0, \tag{3.128}$$

$$\frac{\partial s_0^{(1)}}{\partial t} \frac{\partial s_0^{(2)}}{\partial t} > 0. \tag{3.129}$$

In the case (3.128), the boundaries move in opposite directions. Therefore, since the domains $\Omega_{t,\varepsilon}^{2k}$ or $\Omega_{t,\varepsilon}^{2k+1}$ disappear for $t \sim \varepsilon^\alpha$, the "wave train" structure exists only in a relatively short time interval. A similar situation for the Stefan problem was considered in [38]. For the phase field model, one can prove that, as follows from (3.128), in Ω_t^*, there arises either an "overheating" domain or an "overcooling" domain.

To obtain the conditions under which the "wave train" structure exists independently of ε during a certain time, we consider the case (3.129), where the boundaries move in the same directions. We assume that the following preliminary condition is satisfied.

Assumption 3.13 There exists $T > 0$ such that, for all $t \in [0, T]$, there exist functions $\theta_i(x, t, h)$, $i = 0, \ldots, M + 1$, such that $\tilde{\theta}(x, t, \varepsilon)$ (defined as $\tilde{\theta} = \theta_i$ for $x \in \overline{\Omega^i_{t,\varepsilon}}$) is a continuous function uniformly bounded with respect to $\varepsilon \in [0, \varepsilon_0]$. Moreover, assume that the functions θ_i belong to $C^1(Q^i_\varepsilon)$ uniformly in $\varepsilon \in [0, \varepsilon_0]$ and i, where $Q^i_\varepsilon = \bigcup_{t \in [0,T]} \overline{\Omega^i_{t,\varepsilon}}$, and let $\Gamma^i_{t,\varepsilon} \in C^3$.

Let us consider consequences of this assumption. Obviously, the smoothness of θ_i implies that

$$\theta_i\big|_{\Gamma^i_{t,\varepsilon}-0} - \theta_i\big|_{\Gamma^{i-1}_{t,\varepsilon}+0} = O(h).$$

This fact and the Gibbs–Thomson laws (3.125), (3.126) imply the relation

$$\mathcal{K}_t^i - v_{\mathbf{n}_i} + \mathcal{K}_t^{i-1} - v_{\mathbf{n}_{i-1}} = O(h).$$

But, with regard to our assumptions, the surface $\Gamma_{t,\varepsilon}^i$ is smooth and $v_{\mathbf{n}_{i-1}} v_{\mathbf{n}_i} > 0$. Thus, we have

$$s_0^{(j)}(x, t, h) = s_0(x, t) + h\tilde{s}_0^{(j)}(x, t, h), \quad j = 1, 2, \tag{3.130}$$

where the functions s_0 and $\tilde{s}_0^{(j)}$ and their derivatives up to the third order are uniformly bounded in $h \in [0, h_0]$. Therefore, we obtain

$$v_{\mathbf{n}_i} = \mathcal{K}_t^i + O(h). \tag{3.131}$$

In turn, from (3.131) and (3.125), (3.126) we have

$$\theta_i\big|_{\Gamma_{t,\varepsilon}^i} = O(h).$$

Thus, it follows from Assumption 3.13 that

$$\tilde{\theta}(x, t, \varepsilon) = h\tilde{\theta}^1, \quad \tilde{\theta}^1 = O(1), \quad x \in \overline{\Omega}_{t,\varepsilon}, \quad t \in [0, T], \tag{3.132}$$

where $\tilde{\theta}^1$ is defined as $\tilde{\theta}^1 = \tilde{\theta}_i^1$ for $x \in \overline{\Omega_{t,\varepsilon}^i}$.

Moreover, from the Gibbs–Thomson laws we have

$$\frac{\mathcal{K}_t^i - \mathcal{K}_t^{i-1}}{h} - \frac{v_{\mathbf{n}_i} - v_{\mathbf{n}_{i-1}}}{h} = (-1)^i \kappa_1 \left(\tilde{\theta}_i^1\big|_{\Gamma_{t,\varepsilon}^i - 0} + \tilde{\theta}_i^1\big|_{\Gamma_{t,\varepsilon}^{i-1} + 0} \right).$$

Since $\Gamma_{t,\varepsilon}^i \in C^3$ is uniform in h, we obtain $\tilde{\theta}_i^1 \in C^1(Q_\varepsilon^i)$. For the further analysis, we need the following lemma.

Lemma 3.5 *(a) Let ζ_i be points of partition of the interval $[0, L]$, $\zeta_0 < \zeta_1 < \ldots < \zeta_M$, and let $h = \max\limits_i(\zeta_i - \zeta_{i-1})$. Assume that M is even, $F(\zeta) \in C([0, L])$, and the function $F(\zeta) \in C^1([\zeta_{i-1}, \zeta_i])$ for all $i = 1, \ldots, M$. Then*

$$\left| \sum_{i=0}^{M} (-1)^i F(\zeta_i) \right| \leq \text{const}$$

uniformly in $M \geq 2$.

(b) We assume that $F(\zeta) \in C([0, L])$ and $F(\zeta) \in C^2([\zeta_{i-1}, \zeta_i])$ for all $i = 1, \ldots, M$. Then

$$\sum_{i=0}^{M} (-1)^i F(\zeta_i) = \frac{1}{2}\big(F(\zeta_0) + F(\zeta_M)\big) + O(h)$$

uniformly in even $M \geq 2$.

To prove this lemma, we collect the terms in groups $F(\zeta_i) - F(\zeta_{i-1})$ and represent them as difference derivatives.

We consider the integral identity (3.112). We write

$$\mathcal{T}_\varepsilon^i = \bigcup_{t \in [0,T]} \Gamma_{t,\varepsilon}^i, \quad i = 0, \ldots, M,$$

$$\mathcal{T}_\varepsilon^{M+1} \equiv \mathcal{T}^{M+1} = \partial\Omega \times [0, T],$$

substitute (3.119) into (3.112), and apply Lemma 3.4 to obtain

$$J_\varphi = \sum_{i=0}^{M} \left(\langle g, \nabla s_0 \rangle (\mathcal{K}_t^i - v_{\mathbf{n}_i}) A_{\dot{\chi}^2}, \delta(\mathcal{T}_\varepsilon^i) \right)$$

$$- \sum_{i=0}^{M} (-1)^i \left(\langle g, \nabla s_0 \rangle \theta A_{\dot{\chi}}, \delta(\mathcal{T}_\varepsilon^i) \right) + O(\varepsilon h^{-1} + h) = 0.$$

Applying assertion (a) of Lemma 3.5 to the second sum and using (3.132) and Assumption 3.13, we obtain

$$J_\varphi = \sum_{i=0}^{M} \left(\langle g, \nabla s_0 \rangle (\mathcal{K}_t^i - v_{\mathbf{n}_i}) A_{\dot{\chi}^2}, \delta(\mathcal{T}_\varepsilon^i) \right) + O(\varepsilon h^{-1} + h) = 0. \tag{3.133}$$

Therefore, we again obtain relation (3.131), because, otherwise, the first sum in (3.133) is of the order $O(h^{-1})$.

Taking formula (3.130) into account and passing to the limit as $\varepsilon \to 0$, we see that (3.133) implies that (3.131) is satisfied in the whole domain $\Omega_t^* = \lim_{\varepsilon \to 0} \Omega_{t,\varepsilon}^*$. This means that we obtained the relation

$$|\nabla s_0|^{-1} \frac{\partial s_0}{\partial t} = \operatorname{div} \left(\frac{\nabla s_0}{|\nabla s_0|} \right), \quad x \in \Omega_t^*, \ t > 0.$$

Let us consider the integral identity (3.95). First, we calculate the weak limit of the "right-hand" side $-\varphi_t$ in the heat equation.

Lemma 3.6 *Assume that $\varphi(x, t, \varepsilon) = \varphi_1^{\mathrm{as}}(x, t, \varepsilon) + O(\varepsilon^2)$, where the function φ_1^{as} is defined by formula (3.119), and*

$$c_1 h \leq \operatorname{dist}(\mathcal{T}_\varepsilon^i, \mathcal{T}_\varepsilon^{i+1}) \leq c_2 h, \quad i = 0, \ldots, M - 1,$$

where the constants c_1 and c_2 are independent of ε. Then, for any function $\xi \in C^1(\overline{Q})$ satisfying (3.94), we have

$$\left(\frac{\partial \varphi}{\partial t}, \xi\right) = 2\left(\left\{\sum_{j=0}^{M}(-1)^{j+1} v_{\mathbf{n}_j} \delta(\mathcal{T}_\varepsilon^j)\right\}, \xi\right) + C_1 + O(\varepsilon h^{-1} + h), \qquad (3.134)$$

where $C_1 = O\left(\left(\tilde{s}_0^{(1)} - \tilde{s}_0^{(2)}\right)\big|_{h=0}\right)$ is the contribution of the terms depending on the first approximation of the phase s_0 with respect to h.

We put

$$F(\zeta_k) = \int_{\mathcal{T}_\varepsilon^k} v_{\mathbf{n}_k}\, \mathrm{d}\sigma_k.$$

Then, applying assertion (b) of Lemma 3.5 to (3.134), we obtain

$$\left(\frac{\partial \varphi}{\partial t}, \xi\right) - \int_{\mathcal{T}_\varepsilon^0} \xi v_{\mathbf{n}_0}\, \mathrm{d}\sigma_0 - \int_{\mathcal{T}_\varepsilon^M} \xi v_{\mathbf{n}_M}\, \mathrm{d}\sigma_M + C_1 + O(\varepsilon h^{-1} + h). \qquad (3.135)$$

Recall that, by Assumption 3.13, the family of functions $\left\{\tilde{\theta}(x, t, \varepsilon)\right\}$ is uniformly bounded in ε in $L^\infty\left(0, T; W_2^1(\Omega_t^*)\right)$. Therefore, this family $*$-weakly converges in $L^\infty\left(0, T; W_2^1(\Omega_t^*)\right)$ and, by (3.132), $\lim_{\varepsilon \to 0} \tilde{\theta} = 0$ for $x \in \Omega_t^*$ in the sense of $L^2((0, T) \times \Omega_t^*)$. This implies that

$$\overline{\theta}(x, t) \stackrel{\text{def}}{=} \lim_{\varepsilon \to 0} \tilde{\theta}(x, t, \varepsilon) = 0, \quad x \in \Omega_t^*.$$

It is clear that (3.132) does not contradict (3.135) if and only if the signs of the leading terms in the approximations (depending on $\tilde{s}_0^{(j)}$) of the velocities are independent of j and hence, we have $C_1 = 0$. Otherwise, in the domain Ω_t^*, the limit heat equation has the right-hand side C_1. To verify these facts, we must prove that $\tilde{s}_0^{(1)} = \tilde{s}_0^{(2)} + O(h)$. This will be done later in the spherically symmetric case, because such a geometry is a necessary condition for the existence of the structure in question.

Continuing the calculations of integral identity (3.95), we have

$$\tilde{I}_\theta \stackrel{\text{def}}{=} \int_Q \left\{ -\xi_t \theta + \langle \nabla \xi, \nabla \theta \rangle \right\} \mathrm{d}x \mathrm{d}t + \int_\Omega \xi(x, 0)\theta^0\, \mathrm{d}x.$$

Integrating by parts, we obtain

$$\tilde{I}_\theta = \int\limits_0^T \left\{ \int\limits_{\Omega_{t,\varepsilon}^0} \xi \left(\frac{\partial \theta_{(0)}}{\partial t} - \Delta \theta_{(0)} \right) dx \right.$$

$$+ \int\limits_{\Omega_{t,\varepsilon}^{M+1}} \xi \left(\frac{\partial \theta_{(M+1)}}{\partial t} - \Delta \theta_{(M+1)} \right) dx \bigg\} dt + \int\limits_{\mathcal{T}_\varepsilon^0} \xi \frac{\partial \theta_{(0)}}{\partial \mathbf{n}_0}\, d\sigma_0$$

$$- \int\limits_{\mathcal{T}_\varepsilon^M} \xi \frac{\partial \theta_{(M+1)}}{\partial \mathbf{n}_M}\, d\sigma_M + \sum_{i=1}^M \left\{ \int\limits_0^T \int\limits_{\Omega_{t,\varepsilon}^i} \theta_{(i)} \left(- \frac{\partial \xi}{\partial t} - \Delta \xi \right) dx dt \right.$$

$$+ \int\limits_{\mathcal{T}_\varepsilon^i} \theta_{(i)} \frac{\partial \xi}{\partial \mathbf{n}_i}\, d\sigma_i - \int\limits_{\mathcal{T}_\varepsilon^{i-1}} \theta_{(i)} \frac{\partial \xi}{\partial \mathbf{n}_{i-1}}\, d\sigma_{i-1} \bigg\} + \int\limits_{\Omega_{0,\varepsilon}^*} \xi(x,0) \theta^0\, dx, \quad (3.136)$$

where $\theta_{(i)} = \theta\big|_{Q_\varepsilon^i}$. With (3.132) taken into account, the integrals in (3.136) with respect to $\Omega_{t,\varepsilon}^i$ and $\mathcal{T}_\varepsilon^i$, $i = 1, \ldots, M$, tend to zero as $\varepsilon \to 0$.

Further, it follows from Definition 3.2 that

$$I_\theta = \tilde{I}_\theta - \int\limits_Q \varphi \xi_t\, dx dt + \int\limits_\Omega \varphi^0 \xi(x,0)\, dx = 0.$$

Then, combining (3.131), (3.135), and (3.136) and passing to the limit as $\varepsilon \to 0$, we obtain the desired limiting problem

$$\frac{\partial \overline{\theta}}{\partial t} = \Delta \overline{\theta}, \quad x \in \Omega \setminus \Omega_t^*, \ t > 0; \tag{3.137}$$

$$\overline{\theta} = 0, \quad x \in \Omega_t^*, \ t \geq 0;$$

$$\frac{\partial s_0}{\partial t} = |\nabla s_0|\, \mathrm{div} \left(\frac{\nabla s_0}{|\nabla s_0|} \right), \quad x \in \Omega_t^*, \ t > 0; \tag{3.138}$$

$$\overline{\theta}\bigg|_{\partial \Omega_t^*} = 0, \quad \frac{\partial \overline{\theta}}{\partial \mathbf{n}}\bigg|_{\partial \Omega_t^*} = v_n, \quad t \geq 0; \tag{3.139}$$

$$\overline{\theta}\bigg|_{t=0} = \overline{\theta}^0(x), \quad x \in \Omega \setminus \Omega_0^*;$$

$$s_0\bigg|_{t=0} = s^0(x), \quad x \in \Omega_0^*;$$

$$\overline{\theta}\bigg|_{\partial \Omega} = \theta_b.$$

Here $\partial \Omega_t^* = \Gamma_t^- \cup \Gamma_t^+$, where

$$\Gamma_t^- = \big\{ x \in \Omega, \ s_0(x,t) = 0 \big\}, \quad \Gamma_t^+ = \big\{ x \in \Omega, \ s_0(x,t) = L \big\},$$

n is the outward normal to Ω_t^*,

$$v_{\mathbf{n}} = |\nabla s_0|^{-1} \frac{\partial s_0}{\partial t}\bigg|_{\partial \Omega_t^*},$$

and $s^0(x) \equiv s^0(x, 0)$.

Problem (3.137)–(3.139) can be interpreted as two classical one-phase Stefan problems supplemented with Eq. (3.138). Such an interpretation leads to the statement of the problem with "mushy region" for processes with surface tension [17, 38, 44, 49]. Moreover, conditions (3.138), (3.139) and the relation $\overline{\theta} = 0$ on Ω_t^* are Rankine–Hugoniot-type conditions, because they must be satisfied for the existence of the desired solution.

We make several remarks. The operator in the right-hand side of (3.138) degenerated in the direction ∇s_0, i.e., along y_1, if we introduce new coordinates $y_1 = s_0, y_2, \ldots, y_n$, where $y_2, \ldots, y_n$ are the coordinates on the surface $s_0 = \text{const}$. It is clear that Eq. (3.138) is ultraparabolic. It is well known that homogeneous ultraparabolic equations have no real analytic solutions in t and y_1 except for the case where the solution is independent of the tangential variables [47].

Moreover, for the heat equation (3.137), we must solve the Cauchy problem (3.139) with respect to the variable y_1 with initial conditions on the surface $\partial \Omega_t^*$. For sufficiently small y_1 and t, this ill-posed problem has a solution only for real analytic surfaces and initial conditions [25], and, in this case, the value of $\overline{\theta}$ on the outer boundary and at the initial time is uniquely determined by the value on $\partial \Omega_t^*$. Thus, in particular, for $n = 3$, the solution exists only in the spherically symmetric case with special initial and boundary conditions (on the outer boundary).

Let $n = 3$, $\Omega = \{x, \ R_- < r < R_+\}$, where $r = |x|$, $R_- > 0$, and $\Omega_0^* = \{r_-(0) < r < r_+(0)\}$. Then Eq. (3.138) is the first-order equation

$$\frac{\partial s_0}{\partial t} = \frac{2}{r} \frac{\partial s_0}{\partial r}, \quad r \in \Omega_t^*, \ t > 0, \tag{3.140}$$

where $\Omega_t^* = \{r_-(t) < r < r_+(t)\}$. Obviously, Eq. (3.140) supplemented with the initial conditions $s_0|_{t=0} = s^0(r)$ can be solved easily, i.e., for any smooth function $s^0(r)$ such that $s_r^0 > 0$ along the characteristic $r(r^0, t) = \sqrt{(r^0)^2 - 4t}$,

$$s_0(r, t) = s^0(r^0),$$

where $r_-(0) \leq r^0 \leq r_+(0)$.

Problem (3.137), (3.139), where $v_{\mathbf{n}} = 2/r|_{\partial \Omega_t^*}$, is the Cauchy problem (in the variable r) in the two domains

$$Q_1 = \{R_- < r < r_-(t), t > 0\} \quad \text{and} \quad Q_2 = \{r_+(t) < r < R_+, t > 0\}.$$

To formulate the solvability conditions for this ill-posed problem, we recall the well-known facts (see, e.g., [25, 47]). Namely, for the local existence of solutions of Eqs. (3.137), (3.139), it suffices that the curves $r_{\pm}(t)$ be analytic in t, i.e., $r_-(0) > 0$ and $t < r_-^2(0)/4$. Thus, for sufficiently small $\delta_0 > 0$ and $T_0 = T_0(\delta_0)$, in the domains

$$Q_1^* = \{r_-(0) - \delta_0 < r < r_-(t),\ t < T_0\},$$
$$Q_2^* = \{r_+(t) < r < r_+(0) + \delta_0,\ t < T_0\},$$

there exists a real analytic solution $\overline{\theta}$ of the corresponding Cauchy problem.

Then for the solvability of limiting problem (3.137)–(3.139), we assume that the following condition is satisfied.

Assumption 3.14 Assume that Ω is a spherical layer in $\mathbb{R}^3$, the initial and boundary conditions for problem (3.89), (3.90) are spherically symmetric, and

$$\Gamma_{0,\varepsilon}^i = \left\{ x \in \Omega,\ |x| = r_i^0 \right\},$$

where $0 < R_- < r_0^0 < r_1^0 < \cdots < r_M^0 < R_+$. Assume that $r_{j+1}^0 - r_j^0 = h$, the distances $r_0^0 - R_-$ and $R_+ - r_M^0$ are sufficiently small, the function $s^0(r)$ is real analytic, $\partial s^0/\partial r > 0$, and the functions $\overline{\theta^0}(x)$ and θ_b are special data corresponding to the solution of the Cauchy problem for heat equations (3.137), (3.139).

We return to problem (3.121)–(3.126) and prove that the conditions in Assumption 3.13 are satisfied provided that the conditions in Assumption 3.14 are satisfied and C_1 in (3.135) is zero. By $\rho_i = \rho_i(t, h)$ we denote the functions describing the position of the free boundaries $\Gamma_{t,\varepsilon}^i$ in the time t, i.e., $\Gamma_{t,\varepsilon}^i = \{r, r = \rho_i(t, h)\}$. Clearly, in the spherically symmetric case, $\mathcal{K}_t^i = -2/\rho_i$, and starting from (3.131) and taking the opposite direction (with respect to $D_{t,\varepsilon}^i$) of the normals $\mathbf{n}_i$ into account, we obtain $v_{\mathbf{n}_i} = -2/\rho_i + O(h)$. We change the variable: $\theta_i = w_i/r$. Then Eq. (3.121) becomes

$$\frac{\partial w_i}{\partial t} = \frac{\partial^2 w_i}{\partial r^2}, \quad r \in \big(\rho_{i-1}(t), \rho_i(t)\big),\ t > 0, \tag{3.141}$$

and since $v_{\mathbf{n}_i} = -2\rho_i^{-1}(1 + h\hat{v}_{\mathbf{n}_i})$, where $\hat{v}_{\mathbf{n}_i} \overset{\text{def}}{=} \rho_i(\mathcal{K}_t^i - v_{\mathbf{n}_i})/2h$, conditions (3.122), (3.123), and (3.124) can be rewritten as

$$\frac{\partial w_{i-1}}{\partial r}\bigg|_{\Gamma_{t,\varepsilon}^{i-1}-0} - \frac{\partial w_i}{\partial r}\bigg|_{\Gamma_{t,\varepsilon}^{i-1}+0} = (-1)^i 4(1 + h\hat{v}_{\mathbf{n}_{i-1}}),$$

$$\frac{\partial w_i}{\partial r}\bigg|_{\Gamma_{t,\varepsilon}^i-0} - \frac{\partial w_{i+1}}{\partial r}\bigg|_{\Gamma_{t,\varepsilon}^i+0} = (-1)^{i+1} 4(1 + h\hat{v}_{\mathbf{n}_i}), \tag{3.142}$$

$$w_{i-1}\big|_{\Gamma_{t,\varepsilon}^{i-1}-0} = w_i\big|_{\Gamma_{t,\varepsilon}^{i-1}+0}, \quad w_i\big|_{\Gamma_{t,\varepsilon}^i-0} = w_{i+1}\big|_{\Gamma_{t,\varepsilon}^i+0}.$$

Now we show that the solution of problem (3.141), (3.142) has the following properties.

Property 3.1 The functions $w_i = O(h)$ uniformly in i.

Property 3.2 For all t, the values $\left\{\widehat{w}_i = (-1)^i w_i\big|_{r=\rho_i}\right\}$ with an accuracy of $O(h^2)$ are traces of some functions $\widehat{w} \in C^1[\rho_0, \rho_M]$ on the grid $\{\rho_0, \ldots, \rho_M\}$.

We note that Property 3.1 corresponds to conditions (3.125), (3.126) and relation (3.131).

The solution w_i of problem (3.141), (3.142) can be obtained in the form

$$w_i = a_i(r - \rho_{i-1}) + b_i(t) + u_i(t, r, h), \tag{3.143}$$

where the first two terms correspond to the Stefan conditions (3.142) and u_i are solutions of the system of connected problems:

$$\frac{\partial u_i}{\partial t} - \frac{\partial^2 u_i}{\partial r^2} = a_i \dot{\rho}_{i-1} - \dot{b}_i, \quad i = 1, \ldots, M, \tag{3.144}$$

$$(u_j - u_{j+1})\bigg|_{r=\rho_j} = 0, \quad \left(\frac{\partial u_j}{\partial r} - \frac{\partial u_{j+1}}{\partial r}\right)\bigg|_{r=\rho_j} = 0, \tag{3.145}$$

where $j = 0, \ldots, M$. We note that the given system of problems is similar to that considered in [38]. The difference is that the right-hand sides in (3.144)

$$f_i = a_i \dot{\rho}_{i-1} - \dot{b}_i$$

depend on t, but the contribution to the solution of the system, which ensures this dependence, is of the order $O(h^2)$.

To solve problem (3.144), (3.145), we first calculate the coefficients a_i and b_i. Starting from (3.142) and (3.144), we easily obtain

$$a_i = 2(-1)^{i+1}(1 + h\hat{v}_{\mathbf{n}_i}); \quad b_1 = 0,$$

$$b_j = 2\sum_{k=2}^{j}(-1)^k\left[(1 + h\hat{v}_{\mathbf{n}_{k-1}})\rho_{k-1} - (1 + h\hat{v}_{\mathbf{n}_{k-2}})\rho_{k-2}\right],$$

where $j = 2, \ldots, M$. Assume that a priori

$$\tilde{s}_0^{(j)}(x, t, h) = s_1(x, t) + O(h), \quad j = 1, 2, \tag{3.146}$$

where the functions $\tilde{s}_0^{(j)}$ were defined in (3.130). This assumption can lead to a contradiction in the equation for the velocity correction (the linearized Gibbs–Thomson equation for the functions $\tilde{s}_0^{(j)}$) if the functions ω_i calculated with regard to these conditions do not satisfy Properties 3.1 and 3.2 given above. But, as we can see below, there is no contradiction.

By $R(z, t, h)$ we denote the solution of the implicit equation

$$s_0(R, t) + h s_1(R, t) = z.$$

Then, by construction, $\rho_i = R(ih, t, h)$, and uniformly in i, the functions $\hat{v}_{\mathbf{n}_i}$ are traces of some C^1-functions $\hat{v}$ on the surfaces $r = \rho_i$ with an accuracy of $O(h)$. We note that $\partial R / \partial z > 0$. We have

$$b_j = 2h \sum_{k=2}^{j} (-1)^k \left. \frac{\partial R}{\partial z} \right|_{z=h(k-2)} + O(h^2) = O(h).$$

It follows from Lemma 3.5 that the last estimate is uniform in j. Moreover,

$$b_{j+2} - b_j = 2(-1)^{j+1}(\rho_{j+1} - 2\rho_j + \rho_{j-1}) + O(h^3) = O(h^2), \qquad (3.147)$$

and these estimates are also uniform in j. We note that

$$b_{j+1} - b_j = 2(-1)^{j+1} \left(h \frac{\partial R}{\partial z} + \frac{h^2}{2} \frac{\partial^2 R}{\partial z^2} + h^2 \frac{\partial}{\partial z}(R\hat{v}) \right) \bigg|_{z=(j-1)h} + O(h^3).$$
$$(3.148)$$

Moreover, from (3.147) and Lemma 3.5 we have

$$b_{j+2l} - b_j = O(h^2),$$

and the estimate $O(h^2)$ is uniform in j and l. In particular, this estimate, formula (3.148), and the condition $b_1 = 0$ imply

$$b_{2l} = 2h \left. \frac{\partial R}{\partial z} \right|_{z=(2l-1)h} + O(h^2), \quad b_{2l+1} = O(h^2).$$

We consider a discontinuous line $\mathcal{L}$ whose linear parts are given by the formula $a_i(r - \rho_{i-1}) + b_i$ on the intervals $[\rho_{i-1}, \rho_i]$. Obviously, b_i are the values of $\mathcal{L}$ at the points $r = \rho_{i-1}$. Therefore, the discontinuous line $\mathcal{L}$ is nonsymmetric with respect to the zero line (it can be transformed into a domain with positive values). Thus, the discontinuous function can be centered by decreasing its values on each interval $[\rho_{i-1}, \rho_i]$ by

$$h \left. \frac{\partial R}{\partial z} \right|_{z=h(i-1)}.$$

Clearly, this is equivalent to the subtraction from $\mathcal{L}$ of the function

$$m = h \left. \frac{\partial R}{\partial z} \right|_{z=z(r,t,h)},$$

where $z = z(r, t, h)$ satisfies the implicit equation $R(z, t, h) = r$.

We write $\mathcal{L}_1 = \mathcal{L} - m$ and $U_i = u_i + m$. Then for U_i, we have problems of the form (3.144) with right-hand sides f_i replaced by G_i:

$$G_i = a_i \dot{\rho}_{i-1} - \dot{b}_i + \frac{\partial m}{\partial t} - \frac{\partial^2 m}{\partial r^2}, \quad r \in (\rho_{i-1}, \rho_i).$$

To construct an asymptotic expansion of U_i, we solve our system of problems by a method similar to that in [44]. Namely, we seek the solution in the form

$$\begin{aligned}
U_i &= c_i(r - \rho_{i-1})(\rho_i - r) + c_{i1}(r - \rho_{i-1})^2(\rho_i - r) \\
&\quad + c_{1i}(r - \rho_{i-1})(\rho_i - r)^2 + \cdots,
\end{aligned}$$

where the dots stand for polynomials of higher orders. We note that the polynomials of degree greater than 2 admit the estimate $O(h^3)$ and the coefficients c_i are determined by the relations

$$c_i = 2(-1)^{i+1}\dot{\rho}_{i-1} + O(h), \quad i = 1, \ldots, M.$$

The contribution to the solution U_i of terms $O(h)$ in the right-hand side of G_i is of the order $O(h^3)$.

Thus, $U_i = \widehat{U}_i + O(h^3)$ and $\widehat{U}_i = c_i(r - \rho_{i-1})(\rho_i - r)$ is a sequence of parabolic symmetries $\mathrm{mod}\,O(h^3)$ with respect to zero, because $a_i \dot{\rho}_{i-1} + a_{i+1}\dot{\rho}_i = O(h)$. We note that $\widehat{U}_i = O(h^2)$ for $r \in (\rho_{i-1}, \rho_i)$ and the values of the discontinuous line $\mathcal{L}_1$ at the points ρ_j are given by the formula

$$\mathcal{L}_1\Big|_{r=\rho_j} = (-1)^j h \frac{\partial R}{\partial z}\Big|_{z=(j-1)h} + O(h^2), \tag{3.149}$$

where $j = 1, \ldots, M$. Thus, problem (3.141), (3.142) has a solution satisfying Properties 3.1 and 3.2.

It remains to construct θ in the domains

$$R_- \le r \le \rho_0(t) \quad \text{and} \quad \rho_M(t) \le r \le R_+.$$

We note that the construction of the discontinuous line $\mathcal{L}_1$ allows us to determine the values θ and $\partial\theta/\partial r$ at the points $r = \rho_0(t)$ and $r = \rho_M(t)$ up to $O(h)$. Thus, similarly to (3.137)–(3.139), to complete the construction of θ, we again need to solve the Cauchy problem with respect to r for the heat equation. Necessary conditions for the solvability are satisfied by Assumption 3.14.

Thus, it follows from (3.149) that, with an accuracy of $O(h^2)$, the functions $\hat{\theta}_i(t) = (-1)^i \theta\big|_{r=\rho_i}$ are traces on $\Gamma_{t,\varepsilon}^i$ of the C^1-function $\hat{\theta}(x, t, h) = O(h)$. This fact allows us to calculate the first-order terms of the expansion for the function $s_0(r, t)$. Namely, substituting representation (3.130) in conditions (3.125) and (3.126), we obtain the linearized Gibbs–Thomson conditions

$$\left(\frac{\partial \tilde{s}_0^{(n_i)}}{\partial t} - \frac{2}{r}\frac{\partial \tilde{s}_0^{(n_i)}}{\partial r}\right)\bigg|_{r=\rho_i} = (-1)^i \theta_i \frac{\kappa_1}{h}\frac{\partial s_0}{\partial r}\bigg|_{r=\rho_i} + O(h). \tag{3.150}$$

Our analysis shows that, with an accuracy of $O(h)$, the right-hand side in conditions (3.150) is the trace of a C^1-function. This fact, relation (3.149), and the conditions $\tilde{s}_0^{(1)}\big|_{t=0} = \tilde{s}_0^{(2)}\big|_{t=0} = 0$ imply

$$\left(\frac{\partial s_1}{\partial t} - \frac{2}{r}\frac{\partial s_1}{\partial r}\right)\bigg|_{r=\rho_i} = \frac{\kappa_1}{r}\frac{\partial R}{\partial z}\bigg|_{z=(i-1)h}\frac{\partial s_0}{\partial r}\bigg|_{r=\rho_i} + O(h). \tag{3.151}$$

We note that we can write $\rho_i(t, h) = r_i(t) + h\tilde{r}_i(t, h)$ so that $\tilde{r}_i(t, h)/r_i(t) = O(1)$ uniformly in $i = 0, \ldots, M$. Taking into account Eq. (3.140), we obtain

$$r_i = \sqrt{g^2(ih) - 4t},$$

where $g(z)$ is the inversion of s^0, i.e., $s^0(g(z)) = z$. Thus, omitting the vanishing $O(h)$-terms, we can transform (3.151) as

$$\frac{\partial s_1}{\partial t} = \frac{2}{r}\frac{\partial s_1}{\partial r} + \frac{\kappa_1}{r}, \quad s_1\bigg|_{t=0} = 0.$$

Thus, we obtain a contradiction and hence our assumption about the function $s_1(r, t)$ is satisfied. We again stress that, due to (3.146), $C_1 = 0$ in (3.134) and (3.135), and hence the heat equation has no right-hand side in Ω_t^*. Thus, the fact that Assumption 3.13 is satisfied ensures that Assumption 3.14 is also satisfied. This permits obtaining system (3.137)–(3.139) as the limit system of the Stefan–Gibbs–Thomson problem in the sense of Definition 3.2.

Summarizing the preceding conclusions, we can formulate the following assumptions for the initial conditions.

Assumption 3.15 Assume that the conditions in Assumptions 3.12 and 3.14 are satisfied and $\varphi\big|_{t=0}$ has the form (3.127), where $s^0(x, \varepsilon) = s^0(r)$. Assume also that the function $\theta\big|_{t=0}$ in the domains $\Omega_{0,\varepsilon}^i = \{r_{i-1}^0 < r < r_i^0\}, i = 1, \ldots, M$, is determined by the relations

$$\theta_{(i)}\bigg|_{t=0} = (-1)^{i+1}\frac{2}{r}\left(-\frac{h}{2(s^0)_r'} + (r - r_{i-1}^0) + O(h^2)\right),$$

and in the domains $R_- < r < r_0^0$ and $r_M^0 < r < R_+$, as

$$\theta\big|_{t=0} = \Xi\big|_{t=0},$$

where Ξ is the solution of a special Cauchy problem (with respect to the variable r) for heat equation (3.137).

Fig. 3.3 $t = 0$

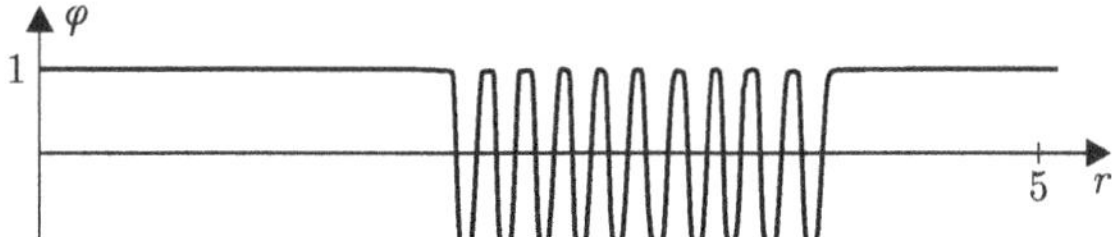

Theorem 3.4 *It follows from the above assumptions that there exists an asymptotic solution of the system of phase field equations satisfying Assumption* 3.13, *and in the system of Eqs.* (3.89), (3.90), *one can pass to the limit as* $\varepsilon \to 0$ *in the sense of Definition* 3.2. *The limiting problem* (3.137)–(3.139) *has a solution at least at small (but ε-independent) times.*

We have considered the case where, outside the stratified domain $r_0 \le r \le r_M$, the order function in the prelimiting problem takes different values: $\varphi \sim -1$ for $r < r_0$ and $\varphi \sim 1$ for $r > r_M$. Obviously, all arguments can be applied in the case where φ takes equal values: $\varphi \sim -1$ or $\varphi \sim 1$ for all $r \notin [r_0, r_M]$. This means that M is odd. Then, obviously, we again obtain the limiting problem in the form (3.137)–(3.139). This passage to the limit can be justified just as above when solving the system of problems (3.121)–(3.126) which can be reduced to the system of problems (3.144).

In both cases (M is even or odd), the problem is ill posed. But surprisingly, solving the phase field system numerically for $R_- \le r \le R_+$ with the initial data $\theta^0 = 0$,

$$\varphi^0 = 1 + \sum_{j=0}^{M}(-1)^j \tanh\left(\frac{r - r_j^0}{\varepsilon}\right)$$

for odd M, and

$$\varphi^0 = \sum_{j-0}^{M}(-1)^j \tanh\left(\frac{r - r_j^0}{\varepsilon}\right)$$

for even M, one observes precisely the "wave train" structure described above.

The graphs shown in Figs. 3.3, 3.4, 3.5, 3.6, 3.7 and 3.8 correspond to the solution of the phase field system in the case of special spherical symmetry and $M = 19$, $\varepsilon = 10^{-2}$ at different times. One can see that the temperature in the mushy region is a sawtooth function, which is only the leading term of asymptotic expansion (3.143) of the solution of Stefan–Gibbs–Thomson problem (3.121)–(3.126). On the outer boundaries, we take $\theta = 0$. This leads to the effect shown in Fig. 3.6: the sawtooth structure of the temperature starts to decay under the action of these nonspecial boundary conditions. At the same time, the order function turns out to be more stable and preserves its shape.

Summarizing this section, we see that the standard definition of weak asymptotic solution is ill posed and does not lead to the Gibbs–Thomson laws even in the usual problem with a fast varying localized perturbation on the free boundary. We have improved this situation by introducing a new definition of weak solution and verifying its well-posedness. When solving the problems that, in general, have no

Fig. 3.4 $t = 0,002$

Fig. 3.5 $t = 0,01$

Fig. 3.6 $t = 0,02$

Fig. 3.7 $t = 0,03$

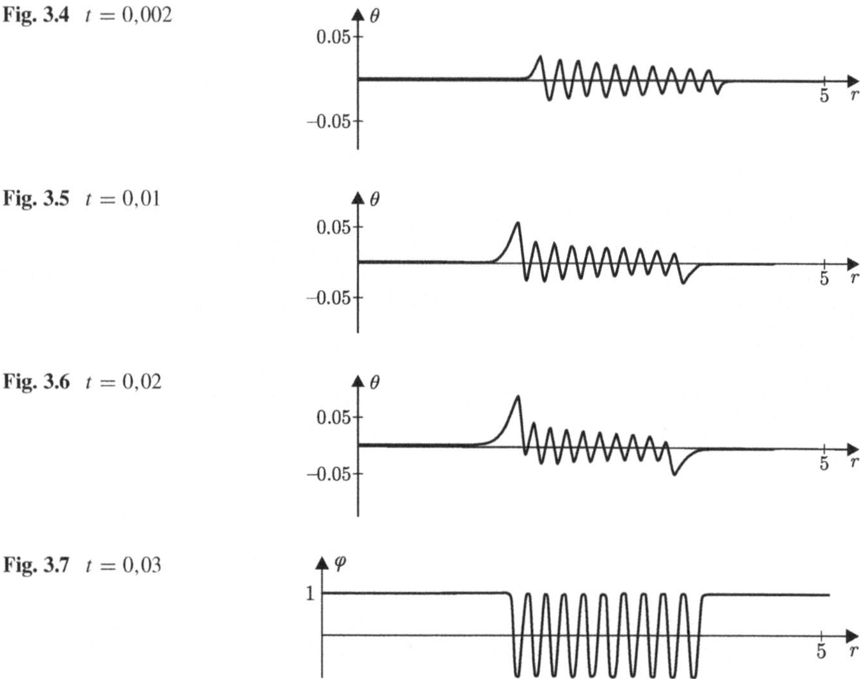

classical solutions, it is very important to have a well-posed definition of weak solution. As an example of such a problem, we considered the well-posed problem with mushy region in the "wave train" interpretation. We proved that a solution of "wave train" type can exist only in the case of special geometry, special boundary conditions, and special initial data. These restrictions ensure the existence of an asymptotic solution of a system of prelimit Stefan–Gibbs–Thomson problems for small (but ε-independent) times. This allows one to pass to the limit in the system of Stefan–Gibbs–Thomson problems in the sense of new Definition 3.2 and to obtain the limiting problem (3.137)–(3.139). We note that, in the situation under study, we discovered that the temperature $\theta(x, t, \varepsilon)$ is small $\left(\lim_{\varepsilon \to 0} \theta = 0\right)$ and has a special "periodic structure" in the mushy region.

In the case of radial symmetry ($n = 2$), the technique described above gives two-dimensional limiting problem (3.137)–(3.139). We note that if, in the original phase field model (3.87), (3.88), the parameters $l \sim k$ are greater, then we obtain heat

Fig. 3.8 $t = 0,03$

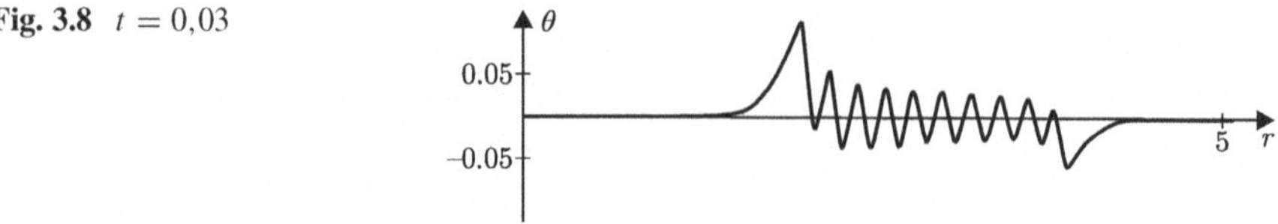

equation (3.87) with $\partial\theta/\partial t$ replaced by $\delta\partial\theta/\partial t$, where $\delta \ll 1$ is a small parameter. Repeating the construction, we obtain heat equation (3.137) with terms $\delta\partial\bar{\theta}/\partial t$ in the right-hand side. Thus, as $\delta \to 0$, Eq. (3.137) turns into the Laplace equation. In this case, the limiting problem (3.137)–(3.139) describes two coupled special Hele-Shaw flows separated by a stagnation region (i.e., "mushy region"). Then the normal velocity $v_{\mathbf{n}}(t)$ of the inner flow satisfies the equation $\dot{v}_{\mathbf{n}} = v_{\mathbf{n}}^{-1}$ on the outer boundary. One can obtain a more general equation for $v_{\mathbf{n}}(t)$ multiplying the Laplace operator in the equation for the order function by a positive real analytic function. Thus, system (3.89), (3.90) formally modified by the above-cited method is a regularization of a class of problems describing the Hele-Shaw flows. This fact permits interpreting the regularization of the Hele-Shaw problem in a way different from [34].

3.5   Derivation of the Solution of the Limit Stefan–Gibbs–Thomson Problem from a Numerical Solution of the Phase Field System

Theorem 3.3 was given in Sect. 3.4.1 mainly to demonstrate what type of the mathematical apparatus we need to use to obtain a rigorous justification of passing to the limit from the regularized problem to the Stefan–Gibbs–Thomson problem under study. In what follows, we do not use this technique.

Nevertheless, in connection with Theorem 3.3, we make several remarks.

Remark 3.2   In fact, the last relations in (3.116) and (3.117) form a definition (somewhat unusual, see [22]) of weak solution of system (3.13).

In particular, (3.117) is the statement in the weak sense of Gibbs–Thomson condition (3.9).

More precisely, Theorem 3.3 given above does not guarantee that the existence of a weak solution of (3.13) (which is stated in Theorem 3.3) necessarily leads to the limit classical solution with the separation of the domain into subdomains filled with the solid and liquid phases. For example, in the previous section, an example of the solution of the Stefan–Gibbs–Thomson problem describing the so-called "mushy region" was constructed. In this region, $\bar{\varphi} \equiv 0$ and $\bar{\theta} \equiv 0$ but $|\nabla\bar{\theta}| =$ const (all relations are understood in the weak sense). Moreover, Theorem 3.3 does not state that even in the case where there exist domains filled with the liquid and solid phases, the boundaries of these domains have at least a certain smoothness. A boundary is the set

$$\Gamma_t = \left\{ x,\ \bar{\varphi}(x,t) = 0 \right\},$$

but $\bar{\varphi} \in BVC$, and hence the set Γ_t may not be a surface at all, despite the fact that, for $\varepsilon > 0$, the regularized phase field system has rather "good" solutions.

Roughly speaking, for the limit of the solution of the phase field system to be indeed the classical solution of the Stefan–Gibbs–Thomson problem, it is required that

$$\Gamma_t^{\varepsilon} = \left\{ x, \; \varphi(x, t, \varepsilon) = 0 \right\}$$

be a surface as $\varepsilon \geq 0$ and the temperature vary slowly in a neighborhood of Γ_t^{ε}, $|\nabla \theta| < \text{const}$ for $\varepsilon \geq 0$.

In numerical computations, one can visually determine whether these conditions are satisfied (or not) for small ε.

Now we return to the method for obtaining the phase field system. From the standpoint of constructing the asymptotic solution for the system of Eq. (3.13) with a small parameter, condition (3.9) is the solvability condition for the equation for the correction to the solution of the second equation (Allen–Cahn equation) in (3.13). Therefore, one can try formally to write another equation (not the Allen–Cahn equation) for which the solvability condition for the equation for the correction is again condition (3.9). It turns out that this can be done, but the simplest version is still the Allen–Cahn equation itself and some small modifications of it. Thus, the system of Eq. (3.13) can be treated in two ways, as a system that has a physical meaning and as a regularization method for the Stefan–Gibbs–Thomson problem obtained from a regularizing problem (for example, problem (3.13)) as $\varepsilon \to 0$. Such a double point of view will be used below.

The phase field system permits significantly simplifying the numerical investigation of the Stefan–Gibbs–Thomson problem. The main advantage is that the equations in the system can be solved in the entire domain while the free boundary can be determined as the support of the jump of the limit function $\bar{\varphi}$.

As was already mentioned, the nonlinearity in the Allen–Cahn equation which has the form

$$g(\varphi) = \varphi(1 - \varphi^2)$$

has three roots $\varphi_0 = 0$ and $\varphi_{\pm} = \pm 1$. By setting $\theta = 0$, one can easily verify that the root φ_0 is unstable and the roots $\varphi_{\pm}$ are stable and the positive perturbations of φ_0 develop into φ_+ while the negative ones develop into φ_-. The solution $\varphi(x, t, \varepsilon)$ is contained in the pair (θ, φ) and has the following structure.

In the ε-neighborhood $\Gamma_{\varepsilon}(t)$ of some smooth surface $\Gamma(t)$, the function φ belongs to $(-1, 1)$, and outside this neighborhood, the function φ is close to the value $\varphi = 1$ or to the value $\varphi = -1$. As $\varepsilon \to 0$, $\Gamma_{\varepsilon}(t)$ becomes the interface $\Gamma(t)$ between phases, the domain

$$\Omega_{\text{sol}} = \left\{ x, \; \bar{\varphi}(x, t) = -1 \right\}$$

turns out to be occupied by a solid phase, and the domain

$$\Omega_{\text{liq}} = \left\{ x, \; \bar{\varphi}(x, t) = 1 \right\}$$

is occupied by the liquid phase.

Considering θ in the Allen–Cahn equation as a perturbing right-hand side, one can understand that, for $\theta < 0$, the values of the function φ rapidly tend to φ_-, and for $\theta > 0$, to φ_+. This allows us to conclude that, in general, the phase field system contains the problem of a two-phase state origination, i.e., the problem of melting or crystallization.

Apparently, for this last problem, the phase field system is only one of the regularization possibilities as follows from the above remarks about the difficulties in determining the domains occupied by phases.

The phase field system has already been used for a long time in computations related to thermal processes with phase transitions, see [35, 39] and the references therein.

It was also noted in [39] that it is very difficult to solve the phase field system numerically if the coefficient at the velocity in Stefan–Gibbs–Thomson conditions (3.9) is significantly less than the other coefficients, i.e., if the temperature on the boundary is mainly influenced by the curvature. In [39], a modification of the phase field system is proposed precisely in this situation. Below we also meet such a situation, but the solution will be different.

We shall consider in more detail the relationship between the solution of the phase field system and the classical solution of the Stefan–Gibbs–Thomson problem under the assumption that the latter exists. We denote this solution by $\bar{\theta}_0$, and assume that the function $\bar{\varphi}_0$ takes the values $+1$ or -1, and $\bar{\theta}_0$ is respectively positive or negative. Assume that the surface of the jump of the function $\bar{\varphi}_0$ coincides with the zero level set of a smooth function $S = S(x, t)$.

For our future plans, it would be better to write (3.13) with some numerical coefficients (undefined at the moment) which will be defined in connection with the physical meaning later in Chap. 4. So let us consider the phase filed system in the form

$$\frac{\partial \theta}{\partial t} + \frac{l}{2} \frac{\partial \varphi}{\partial t} - k\Delta\theta, \tag{3.152}$$

$$\varepsilon^2 \bar{\alpha} \frac{\partial \varphi}{\partial t} - \varepsilon^2 \bar{\beta}^2 \Delta\varphi - \varphi(1 - \varphi^2) = \frac{\overline{\Delta s}}{2} \varepsilon\theta. \tag{3.153}$$

As was shown above, in our case, there exists a pair of functions $\theta^{\mathrm{as}}(x, t, \varepsilon)$, $\varphi^{\mathrm{as}}(x, t, \varepsilon)$ of the form

$$\theta^{\mathrm{as}} \cong \bar{\theta}^-(x, t) + \big(\bar{\theta}^+(x, t) - \bar{\theta}^-(x, t)\big)\omega_1\left(\frac{S}{\varepsilon|\nabla S|}\right), \tag{3.154}$$

$$\varphi^{\mathrm{as}} \cong \omega_0\left(\frac{S}{\varepsilon|\nabla S|}\right). \tag{3.155}$$

Here $\lim\limits_{z\to\infty} \omega_1(z) = 0$, $\lim\limits_{z\to-\infty} \omega_1(z) = 1$, where $\omega_1^{(k)}(z) \in \mathcal{S}(\mathbb{R}^1_z)$ for $k > 0$, $\omega_0(z) = \tanh\left(z/\sqrt{2}\right)$, and $\mathcal{S}(\mathbb{R}^n_z)$ is the Schwartz space of smooth rapidly decreasing functions. In this case, the functions in the right hand sides of (3.154) and (3.155) are the leading terms of asymptotic solutions θ^{as}, φ^{as} satisfying Eqs. (3.152), (3.153) with an arbitrary accuracy in ε (in formulas (3.154) and (3.155), we write only the leading terms of the asymptotics, see Sect. 3.3 for details).

Moreover, one can show that, in this case, there exists an exact solution of system (3.152), (3.153) which arbitrarily small (up to $O(\varepsilon^N)$) differs from the asymptotic solution (3.154), (3.155), see [16].

The converse assertion is unknown but one can formulate an algorithm for using numerically constructed solutions of system (3.152), (3.153) to establish the correspondence between them and the solutions of the Stefan–Gibbs–Thomson problem.

We distinguish the numerical solutions of system (3.152), (3.153) and its exact solutions, because the numerical solutions satisfy the difference equations approximating differential equations (3.152), (3.153) rather than these equations. It is necessary to take this difference into account because, as follows from formulas (3.154), (3.155), the solutions of phase field system (3.152), (3.153) corresponding to the classical solutions of the Stefan–Gibbs–Thomson problem vary fast in a neighborhood of the phase transition $\{x;\ S = 0\}$:

$$\frac{\partial \varphi^{as}}{\partial t} \approx \varepsilon^{-1} \quad \text{and} \quad \left|\nabla \varphi^{as}\right| \approx \varepsilon^{-1}.$$

This means that, for small ε, the difference scheme do not give good approximations of Eqs. (3.152), (3.153) in this neighborhood. It is necessary to choose sufficiently small values of ε due to the above-cited specific nature of the problem, i.e., the solutions of system (3.152), (3.153) pass into the solutions of the Stefan–Gibbs–Thomson problem precisely as $\varepsilon \to 0$.

Another problem cited above is also close to this correspondence. In the case where the dynamical coefficient in the Gibbs–Thomson condition is small (α is small in (3.9)), it may in general happen that the numerical solution does not correspond to the solution of the Stefan–Gibbs–Thomson problem because of the errors related to the approximation of the derivative $\partial/\partial t$ and a finite value of ε in the actual computations.

Thus, as was already noted in Sect. 3.4, the second method for establishing the correspondence between the solution of the phase field system and the classical solution of the Stefan–Gibbs–Thomson problem consists in determining the weak solution of the original problem which admits passing to the limit in the weak sense.

We stress that the problem of determining the weak solution is not specific precisely in the nonlinear case. As is known, for linear equations with smooth dependence of the coefficients on the variables, there is a standard method for determining the weak solution. Namely, it consists in multiplying the equation by a test function and transferring the derivatives to this function (integrating by parts). But in the nonlinear case, such an operation does not lead to a correct definition. This situation is described in detail in Sect. 3.4 (also see [20, 22]) and we do not consider it in detail but simply use the definition constructed there.

In the case of phase field system (3.152), (3.153), the first equation (of heat conduction) is linear, and hence we mainly concentrate on the second (Allen–Cahn) equation.

The definition of weak asymptotic solution for phase field system (3.152), (3.153) has the form of integral identities. For the heat equation, this identity is standard and is given by the formula

$$\int_{\Omega} \theta_t \zeta \, \mathrm{d}x \mathrm{d}t + \frac{l}{2} \int_{\Omega} \varphi_t \zeta \, \mathrm{d}x \mathrm{d}t - k \int_{\Omega} \Delta \theta \zeta \, \mathrm{d}x \mathrm{d}t = \widetilde{R}_{\varepsilon},$$

where $\zeta = \zeta(x, t)$ is a test function and $\widetilde{R}_{\varepsilon} \to 0$ as $\varepsilon \to 0$. For the second (Allen–Cahn) equation in system (3.152), (3.153), the definition of the weak solution has the form

$$\bar{\alpha}\varepsilon \int_{Q} \varphi_t \langle \chi, \nabla\varphi \rangle \mathrm{d}x \mathrm{d}t + \frac{\bar{\beta}^2 \varepsilon}{2} \int_{Q} \left[|\nabla\varphi|^2 + \frac{1}{2\varepsilon\bar{\beta}^2}\left(\varphi^2 - 1\right)^2 \right] \mathrm{div} \ \chi \, \mathrm{d}x \mathrm{d}t$$

$$+ \varepsilon\bar{\beta}^2 \int_{Q} \left[\left\langle \nabla\varphi, \frac{\partial\chi}{\partial x}\nabla\varphi \right\rangle + \frac{\overline{\Delta s}}{2\bar{\beta}^2 \varepsilon}\varphi \, \mathrm{div} \ (\chi\theta) \right] \mathrm{d}x \mathrm{d}t = R_{\varepsilon}, \qquad (3.156)$$

which must be satisfied for any test vector function $\chi(x, t)$, R_{ε} is a certain quantity, and $R_{\varepsilon} \to 0$ as $\varepsilon \to 0$, see Sect. 3.4 and [20, 23].

As was noted above, see Sect. 3.4, relation (3.156) is obtained from equation (3.153) by multiplying by the expression $\langle \chi, \nabla\varphi \rangle$ and subsequently integrating by parts. In the case of this multiplication, we obtain expressions that have no divergence form, and therefore, not all derivatives can be transferred to the test function. This distinguishes the proposed construction (3.156) from the scheme common in the theory of linear differential equations. The usual construction gives an "ill-posed" definition that does not permit passing to the limit. This is explained in detail in Sect. 3.4.

Now we show how one can pass to the weak limit in (3.156). As was previously noted, there do not exist conditions ensuring the existence of such a solution if the existence of the classical solution of the limiting problem is a priori unknown, and to determine this solution is precisely our goal! This seems to be impossible, but the use of computers can help in this case.

The graphs constructed from the results of computations (see Sect. 4.7) clearly show the structure of the solution φ corresponding to the classical solution, see Figs. 4.9a and 4.11 in Sect. 4.7, respectively. This must be a smoothed step for which the passage from the values ± 1 occurs in a small neighborhood of a certain surface, i.e.,

$$\varphi(x, t, \varepsilon) = \omega\left(\frac{S}{\varepsilon|\nabla S|\hat{\beta}}\right) + O(\varepsilon), \qquad (3.157)$$

where S is a smooth function such that $\nabla S\big|_{s=0} \neq 0$, $\omega(z) \in \mathbf{C}^\infty$ is such that $\omega(+\infty) = 1$ and $\omega(-\infty) = -1$, $d^\alpha \omega/dz^\alpha = o\big(|z|^{-N}\big)$ as $|z| \to \infty$, and $N > 0$ is a sufficiently large number. The condition $\nabla S\big|_{s=0} \neq 0$ ensures that the set $\{x;\ S(x,t) = 0\}$ is a smooth surface. Formula (3.157) means that, for $S > 0$, the ratio S/ε tends to $+\infty$ as $\varepsilon \to +0$ and $\omega(S/\varepsilon) \to +1$. In turn, for $S < 0$, $S/\varepsilon \to -\infty$ and $\omega(S/\varepsilon) \to -1$. The term $O(\varepsilon)$ in (3.157) requires a certain detailed procedure which we omit. We simply assume that it is small and insignificant. The further reasoning is based on Lemma 3.4. Now we note that

$$\frac{\partial}{\partial t}\omega\left(\frac{S}{\varepsilon\hat{\beta}|\nabla S|}\right) = \frac{S_t}{\varepsilon\hat{\beta}|\nabla S|}\dot{\omega}\left(\frac{S}{\varepsilon\hat{\beta}|\nabla S|}\right) + \frac{S}{\varepsilon\hat{\beta}}\dot{\omega}\left(\frac{S}{\varepsilon\hat{\beta}|\nabla S|}\right)\frac{\partial}{\partial t}\frac{1}{|\nabla S|}.$$

The last term can be written as

$$\frac{S}{\varepsilon\hat{\beta}}\dot{\omega}\left(\frac{S}{\varepsilon\hat{\beta}|\nabla S|}\right)\frac{\partial}{\partial t}\frac{1}{|\nabla S|} = z\left[|\nabla S|\frac{\partial}{\partial t}\frac{1}{|\nabla S|}\right]\dot{\omega}(z)\Bigg|_{z=\frac{S}{\varepsilon\hat{\beta}|\nabla S|}}.$$

We see that the right-hand side of the last relation is bounded as $\varepsilon \to 0$ by the conditions imposed on the function ω. Now we apply Lemma 3.4 to write

$$\frac{\partial}{\partial t}\omega\left(\frac{S}{\varepsilon\hat{\beta}|\nabla S|}\right) = 2\frac{S_t}{\hat{\beta}|\nabla S|}\delta\left(\frac{S}{\hat{\beta}|\nabla S|}\right) + O_{\mathcal{D}'}(\varepsilon^2), \qquad (3.158)$$

where $O_{\mathcal{D}'}(\varepsilon^2)$ is a generalized function whose action on any test function $\chi(x)$ is estimated as $O(\varepsilon^2)$,

$$\left\langle O_{\mathcal{D}'}(\varepsilon^2), \chi \right\rangle = O(\varepsilon^2)$$

for any $\chi \in \mathbf{C}_0^\infty(Q)$.

Now we note that the expression $\big|S_t/|\nabla S|\big|$ is the absolute value of the velocity of a point on the surface $\{x;\ S = 0\}$ in the direction of the normal on this surface. Indeed,

$$S\big(x_1(t), x_2(t), \ldots, x_n(t), t\big) = 0 \qquad (3.159)$$

if the point $\big(x_1(t), x_2(t), \ldots, x_n(t)\big)$ belongs to $\{x;\ S(x,t) = 0\}$. We differentiate (3.159) with respect to t to obtain

$$S_t + \big\langle \nabla S, \dot{x} \big\rangle = 0$$

or

$$v_\mathbf{n} = \left\langle \frac{\nabla S}{|\nabla S|}, \dot{x} \right\rangle = -\frac{S_t}{|\nabla S|}.$$

Now if we calculate all terms in the left-hand side of (3.156) similarly to (3.158), then we obtain

$$
\bar{\alpha}\varepsilon \int_Q \varphi_t \langle \chi, \nabla\varphi\rangle \mathrm{d}x\mathrm{d}t + \frac{\bar{\beta}^2 \varepsilon}{2} \int_Q \left[|\nabla\varphi|^2 + \frac{1}{2\varepsilon\bar{\beta}^2}(\varphi^2 - 1)^2 \right] \operatorname{div} \chi \, \mathrm{d}x\mathrm{d}t
$$

$$
+ \varepsilon\bar{\beta}^2 \int_Q \left[\left\langle \nabla\varphi, \frac{\partial\chi}{\partial x}\nabla\varphi \right\rangle + \frac{\overline{\Delta s}}{2\bar{\beta}^2\varepsilon}\varphi \operatorname{div}(\chi\theta) \right] \mathrm{d}x\mathrm{d}t = \int_{S=0} \left[-\frac{\bar{\alpha}}{\hat{\beta}}\sigma_1(v_{\mathbf{n}} + O(\varepsilon)) \right.
$$

$$
\left. -\sigma_1\bar{\beta}\big(K + O(\varepsilon)\big) - \overline{\Delta s}\big(\theta + O(\varepsilon)\big) \right] \langle \chi, \mathbf{n}\rangle \mathrm{d}\sigma + \sigma_2, \qquad (3.160)
$$

where (see Lemma 3.4) $\mathrm{d}\sigma$ is the Leray measure on $\{S = 0\}$, $\bar{\alpha}, \bar{\beta}$ and $\overline{\Delta s}$ are constant constants without any physical meaning, and

$$
\sigma_1 = \int_{-\infty}^{\infty} \dot{\omega}^2(z)\,\mathrm{d}z, \quad \sigma_2 = \int_{S=0}\int_{-\infty}^{\infty} z\big(\ddot{\omega}(z) + \omega(z)(1 - \omega^2(z))\big)\dot{\omega}(z)G(x,t)\,\mathrm{d}z\mathrm{d}\sigma.
$$

Here we took into account that, for each odd function ω,

$$
\int_{-\infty}^{\infty} \big[\ddot{\omega}(z) + \omega(z)(1 - \omega^2(z))\big]\dot{\omega}(z)\,\mathrm{d}z = 0,
$$

and G is a function constructed from χ, its derivatives, and the derivatives of S. Of course, if (at the moment) we do not use the regular asymptotics, then the exact formula for $\omega(\tau)$ is unknown. But it is clear that $\Omega(\tau)$ in exact form enters only the coefficient σ_1 as integral which can easily be calculated using the result of numerical calculations.

The structure of the right-hand side of (3.160) can be determined from the above calculations (see, e.g., (3.158)), and it is very important. This means that the phase field approximates the Stefan–Gibbs–Thomson problem even in the case of small coefficients in the Gibbs–Thomson condition. Indeed, if, for example, instead of the term $-\bar{\alpha}\big(v_{\mathbf{n}} + O(\varepsilon)\big)$ in the right-hand side of (3.160), we had $-\bar{\alpha}v_{\mathbf{n}} + O(\varepsilon)$, then in the case of small $\bar{\alpha}$ and the values of ε that are "not close" to zero, the correction $O(\varepsilon)$ could play a role more significant than the physically meaningful term $-\bar{\alpha}v_{\mathbf{n}}$.

If we add the right-hand side of (3.156) to (3.160), then we can formulate the following nonformal assertion.

Proposition 3.1 *Assume that a function of the order $\varphi(x, t, \varepsilon)$ has the form* (3.157) *and the temperature θ has no jump in the domain of fast variation of the function $\varphi(x, t, \varepsilon)$. Then, if $\overline{\Delta s} = \sigma_1 \Delta s$ together with the relation*

$$
\int_{-\infty}^{\infty} z\big(\ddot{\omega}(z) + \omega(z)(1 - \omega^2(z))\big)\dot{\omega}(z)\,\mathrm{d}z = 0,
$$

the following relation holds:

$$\left[-\frac{\bar{\alpha}}{\bar{\beta}}(v_{\mathbf{n}} + O(\varepsilon)) - \bar{\beta}(K + O(\varepsilon)) - \Delta s(\theta + O(\varepsilon)) \right]\Big|_{\{S=0\}} = R_\varepsilon\Big|_{\{S=0\}}.$$

The relation mentioned in Proposition 3.1 obviously holds if $\omega(z)$ is the solution of the problem

$$\ddot{\omega}(z) + \omega(z)(1 - \omega^2(z)) = 0, \qquad \omega(-\infty) = -1, \quad \omega(\infty) = 1.$$

It remains to understand the cause of possible appearance of the quantity R_ε and to estimate it.

The first question has an easy answer: using computer to solve the problem, we do not obtain the exact value of a solution, and substituting the numerical solution into the differential equation, we obtain a discrepancy due to the error of approximation of the differential operator by the difference one.

Recall that if the difference approximation of the operator d/dx has the form

$$u_x = \pm \frac{u(x \pm h) - u(x)}{h},$$

then the approximation error is proportional to the expression

$$h\frac{\partial^2 u}{\partial x^2},$$

where h is the step of the grid. In the case under study, $u = \omega(S/\varepsilon)$ and the differentiation results in the appearance of negative powers of ε as multipliers, and hence it is necessary to estimate the approximation error.

We consider only the one-dimensional case, because the structure of the approximation error does not change depending on the dimension. It is easy to see that the error of approximation of the first term in the left-hand side of (3.156) consists of two terms:

$$I_1 = \frac{\tau \bar{\alpha} \varepsilon}{\varepsilon^3 \bar{\beta}^3} \int_Q \dot{\omega}\left(\frac{S}{\varepsilon \bar{\beta}|\nabla S|}\right)\ddot{\omega}\left(\frac{S}{\varepsilon \bar{\beta}|\nabla S|}\right)g_1(x, t)\,dxdt, \tag{3.161}$$

$$I_2 = \frac{h^2 \bar{\beta}^2 \varepsilon}{\varepsilon^5 \bar{\beta}^5} \int_Q \dot{\omega}\left(\frac{S}{\varepsilon \bar{\beta}|\nabla S|}\right)\omega^{(4)}\left(\frac{S}{\varepsilon \bar{\beta}|\nabla S|}\right)g_2(x, t)\,dxdt, \tag{3.162}$$

where $g_1(x, t)$ and $g_2(x, t)$ are smooth functions composed of the test function χ and the derivatives of the function S; the dot denotes the derivative $d\omega/dz = \dot{\omega}(z)$.

Using the change of variables as in the proof of Lemma 3.4 in Sect. 3.4, we easily verify that expression (3.161) can be estimated by the quantity

$$I_1 = O\left(\frac{\bar{\alpha}\tau}{\varepsilon\bar{\beta}^2}\right),$$

and expression (3.162), by the quantity

$$I_2 = O\left(\frac{h^2}{\varepsilon^3\bar{\beta}^2}\right).$$

In the second estimate, we take into account that $\int \ddot{\omega}(z)\ddot{\omega}(z)\mathrm{d}z = 0$.

The other terms in (3.156) can be considered similarly. This leads to the following expression for R_ε where we took into account that φ in this case is an exact solution to the difference equation. If we substitute the solution of the difference equation into the left-hand side instead of $\varphi(x, t, \varepsilon)$, then

$$R_\varepsilon = O\left(\frac{\bar{\alpha}\tau}{\varepsilon\bar{\beta}^2}\right) + O\left(\frac{h^2}{\varepsilon^3\bar{\beta}^2}\right). \tag{3.163}$$

Thus, we obtain the following assertion.

Proposition 3.2 *Assume that the numerical simulation gives a solution that satisfies the conditions of Proposition* 3.1. *Then*

$$\left[-\frac{\bar{\alpha}}{\bar{\beta}}\left(v_{\mathbf{n}} + O(\varepsilon)\right) - \bar{\beta}\left(K + O(\varepsilon)\right) - \Delta s\left(\theta + O(\varepsilon)\right)\right]\Bigg|_{\{S=0\}} = R_\varepsilon\Bigg|_{\{S=0\}}.$$

where the remainder R_ε is determined by formula (3.163).

We arrive at the following conclusions.

First, independently of the choice of the values of the coefficients of the Allen–Cahn equation in phase field system (3.152), (3.153), the definition of the weak asymptotic solution (3.156) gives a correct result.

Second, solving the difference problem exactly, we obtain Gibbs–Thomson condition (3.9) (which is obtained by passing to the limit from the Allen–Cahn equation) with appropriate corrections. In other words, if we specify the values of the coefficients in the Allen–Cahn equation, then we obtain a numerical solution of the Stefan–Gibbs–Thomson problem with appropriate values of the coefficients. The classical solution of the Stefan–Gibbs–Thomson problem can easily be identified from the graph of the numerical solution, see Figs. 4.9a and 4.11 in Sect. 4.7.

3.6   Generation and Coalescence of Dissipative Waves

In Sect. 3.4, we describe the process of mushy region generation. As was said above, this object corresponds to the nonclassical solution of Stefan–Gibbs–Thomson problem related to the order function whose limit as $\varepsilon \to 0$ does not belong to $BVC(\Omega)$.

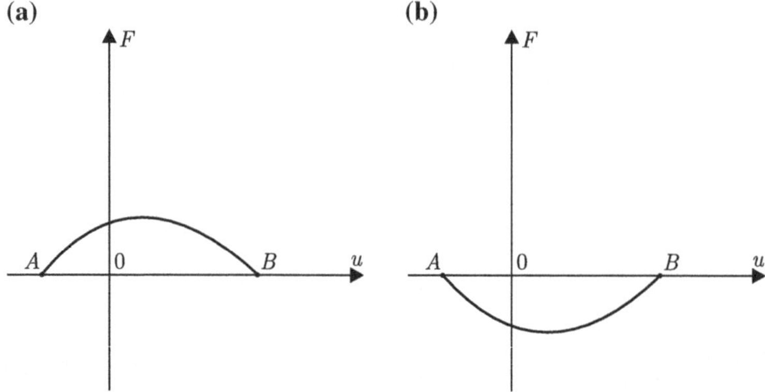

Fig. 3.9 Typical nonlinearities with two roots. Nonlinearity is positive (**a**) or negative (**b**) between roots

The construction of this object is closely related to a sequence of noninteracting nonlinear waves. In what follows, we will use another object related to the beginning of the process of melting.

We call this procedure the introduction of a liquid phase nucleus. This nucleus finally becomes a domain filled by liquid phase, and from the nonlinear interaction viewpoint, this is the coalescence of nonlinear waves but backward in time, i.e., the process of nonlinear waves creation. We explain these processes at a qualitative level in the framework of semilinear parabolic equation and its solutions of nonlinear wave type.

Recently,[2] new methods for integrating nonlinear equations with sufficiently many conservation laws were proposed. But these methods cannot be used to integrate semilinear parabolic equations describing dissipative processes. For such equations, only explicit formulas for self-similar solutions have been known well enough, i.e., formulas for the solutions $u(x, t)$ that can be represented in the form

$$u(x, t) = \chi(\tau), \tag{3.164}$$

where τ is a self-similar variable, $\tau = x + pt$, and $p = \text{const}$.

First, solutions of semilinear parabolic equations were studied in the already classical works by Kolmogorov et al. [37], where they considered the self-similar solutions of the equation

$$u_t - u_{xx} - F(u) = 0, \tag{3.165}$$

choosing the values between the roots of the equation $F(u) = 0$ (we denote them by A and B, $A < B$, see Fig. 3.9, under the assumption that $F(A) = F(B) = 0$ and $F'(A) \cdot F'(B) < 0$). In this case, the self-similar solution is defined by the solution of the equation

---

[2]For more details about the problems considered here and references to the literature, see [19].

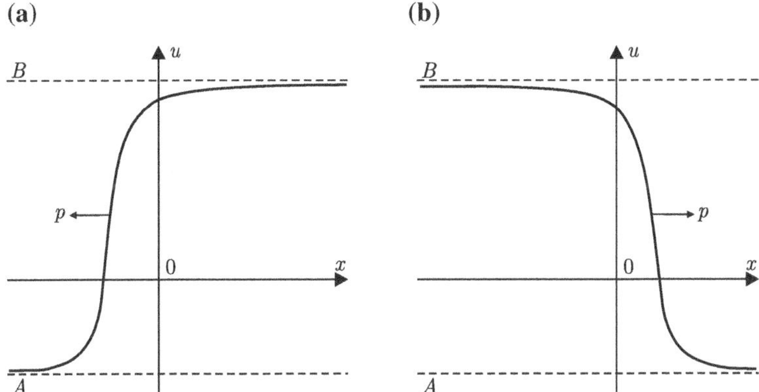

Fig. 3.10 Simple waves corresponding to the nonlinearity in Fig. 3.9a

$$px' - \chi'' - F(\chi) = 0, \tag{3.166}$$

which satisfied the boundary conditions

$$\chi\big|_{\tau \to -\infty} \to A, \quad \chi\big|_{\tau \to +\infty} \to B \tag{3.167}$$

or

$$\chi\big|_{\tau \to -\infty} \to B, \quad \chi\big|_{\tau \to +\infty} \to A. \tag{3.168}$$

It was proved in [37, 56] that a solution of problem (3.166)–(3.168) exists for $p \geq p_{\min} = 2\sqrt{F'(A)}$ (for the case $F'(A) > 0$, see Fig. 3.9a). Solution (3.164) of problem (3.166), (3.167) has the form shown in Fig. 3.10a. Since Eq. (3.165) is invariant under the change of variables $x \to -x$, problem (3.165), (3.168) also has wave solutions shown in Fig. 3.10b.

For $F'(B) > 0$ (the function $F(u)$ changes sign, see Fig. 3.9b), the solution of problem (3.166)–(3.168) exists for $p \geq p_{\min} = 2\sqrt{F'(B)}$ and the wave solutions of problems (3.165), (3.167) and (3.165), (3.168) have the form shown in Fig. 3.11a, b, respectively.

Assume that the equation $F(u) = 0$ has three roots (see Fig. 3.12) $F(A) = F(B) = F(C) = 0$. In this case, the graph of the function $F(u)$ is a union of the graphs of the functions $F(u)$ shown in Fig. 3.9. We shall consider the function $F(u)$ shown in Fig. 3.12. Then problems (3.165), (3.167) and (3.165), (3.168) have wave solutions of the form (3.164) represented in Fig. 3.13. These are two types of wave interaction, namely, decay of waves in a strip , see Fig. 3.14a, b, and coalescence of waves propagating in different strips, see Fig. 3.15a, b. Obviously, to construct explicit formulas, it is first necessary to construct explicit formulas describing the propagation of the unit wave. For the simplest Kolmogorov–Petrovskii–Piskunov (KPP) equation

$$u_t - u_{xx} - u(1 - u) = 0, \tag{3.169}$$

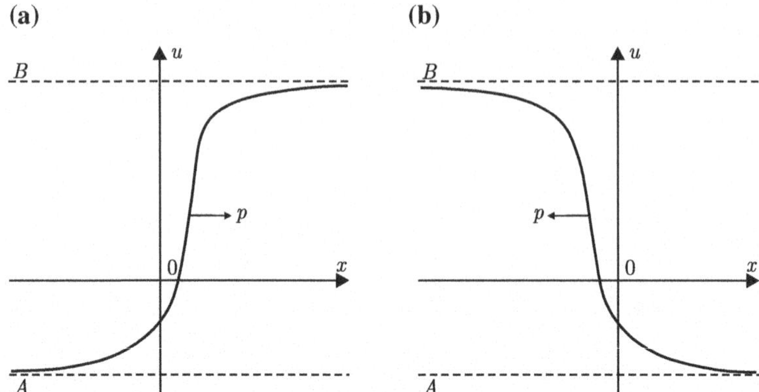

Fig. 3.11 Simple waves corresponding to the nonlinearity in Fig. 3.9b

Fig. 3.12 Nonlinearity with
three roots

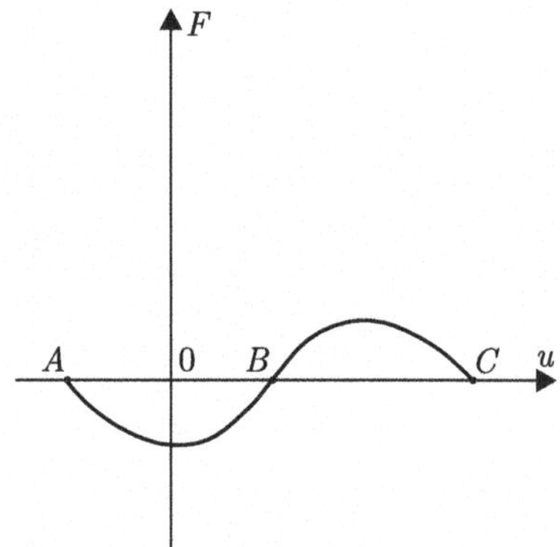

the well-known formula describing the self-similar wave has the form

$$u = \left(1 + e^{-\tau/\sqrt{6}}\right)^{-2}, \quad \tau = x + pt, \quad p = 5/\sqrt{6} > p_{\min} = 2. \tag{3.170}$$

For the equation with cubic nonlinearity, the formulas describing the coalescence-type interaction of waves were obtained in [36] for

$$F = u(1 - u^2)$$

and

$$F = u(1 - u)(v + u).$$

There is a formula in [19] that describes the wave interaction of this type for the equations whose nonlinearity is given by a cubic polynomial $F(u)$ with a negative coefficient at the highest-order term. It should be noted that all formulas given above

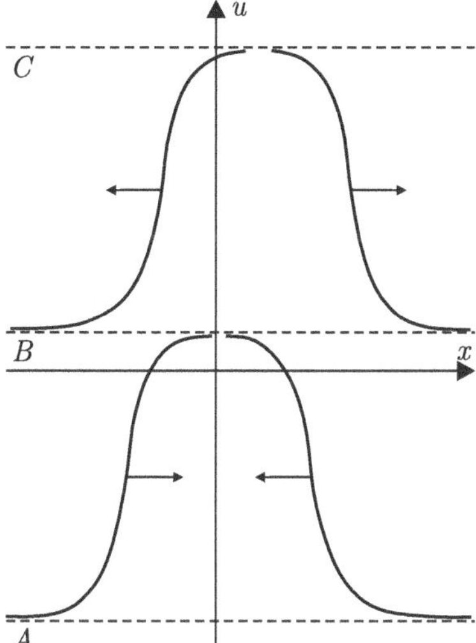

Fig. 3.13 Possible simple waves corresponding to the nonlinearity with three roots

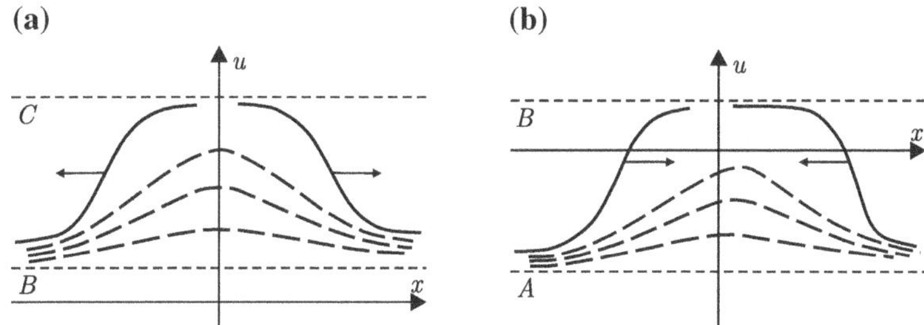

Fig. 3.14 Generation (**a**) and decay (**b**) of simple waves

that describe the wave interaction of coalescence type were constructed under the assumption that each of the interacting waves propagates at a certain definite velocity, i.e., for each equation, there is a known explicit formula describing the coalescence of two self-similar waves with certain definite velocities. This can be explained by the fact that, for Eq. (3.169), only one of the existing waves, which has the velocity $p = 5/\sqrt{6}$, is given by explicit formula (3.170). In [1], the same velocity was obtained by applying the Painlevé test to Eq. (3.169). Moreover, the Painlevé test applied to an equation with nonlinearity in the form of a cubic polynomial gives three possible values of the velocity of the pole motion, and the formulas describing the coalescence were constructed precisely for the waves propagating with these velocities.

It follow from [19, 24] that this property is related to the expansion of the solution in a Dirichlet series. It is well known that the solution of an ordinary differential equation for a self-similar wave can be represented as a Dirichlet series in a neighborhood of the roots of nonlinearity, i.e., in the domain, where the equation can be

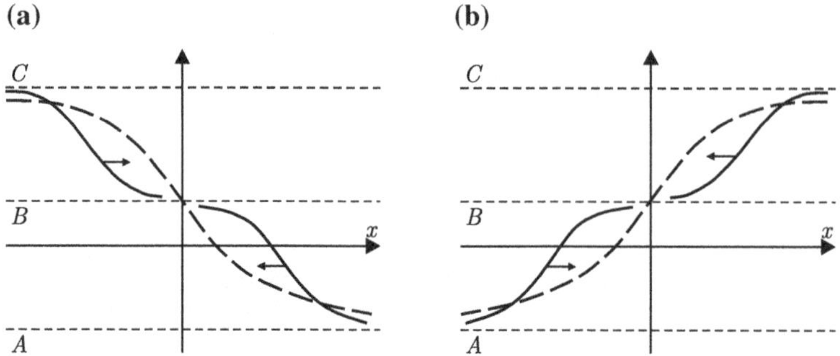

Fig. 3.15 Two types of formation of simple waves from simple waves with values between two roots, see Figs. 3.13 and 4.34, 4.35, 4.36, 4.37, 4.38 as well

linearized "in the main". Assume that we have a series consisting of the terms $e^{\lambda\tau}$ and $e^{-\lambda\tau}$ as $\tau \to -\infty$ and $\tau \to \infty$, respectively, where $\lambda \in \mathbb{R}_+$ and $\tau = x + pt$. The set of such solutions for each equation is very poor [19, 24]. For example, for Eq. (3.169), the constants λ and p are defined in a natural way.

On the other hand, the structure of the Dirichlet series is suggested by the following representation of the self-similar solution:

$$\chi(\tau) = U\left(\frac{\varphi(e^{\lambda\tau})}{\psi(e^{\lambda\tau})}\right), \tag{3.171}$$

where $U = U(z)$ is an analytic function holomorphic near the real axis and at infinity and $\varphi(z)$ and $\psi(z)$ are polynomials. Obviously, representation (3.171) is not unique, but the wave velocity is determine uniquely. Theorem 2.1 in [19] implies a uniqueness criterion for such solutions (up to translations). Further, we note that representation (3.171) is closely related to the results obtained by the Painlevé test, because the solution poles obviously coincide with the set of zeros of the polynomial $\psi(e^{\lambda\tau}) = \psi(e^{\lambda(x+pt)})$ and move with the velocity $\dot{x} = -p$. Our restriction on the search of solutions in the form (3.171) is essential. For example, for Eq. (3.169), the self-similar wave propagating with the minimal velocity cannot be represented in this form. An advantage is that we can use constructive methods to obtain solutions in the form (3.171), for example, the Hirota method and its modifications.

The interaction (coalescence) of waves is described by self-similar solutions of the form (3.171), where polynomials in one variable are replaced by polynomials in two variables. Similar solutions describe the interaction of solitons and kinks for equations of sine-Gordon-type. Apparently, there exists a common, still unknown, mechanism that guarantees the existence of such solutions for both integrable and nonintegrable equations. This hypothesis is confirmed in [13, 58].

A separate problem is to construct an explicit formula describing the interaction of waves of creation or annihilation type. We write an explicit formula only for the leading term of the asymptotic solution describing the generation (decay) of a wave

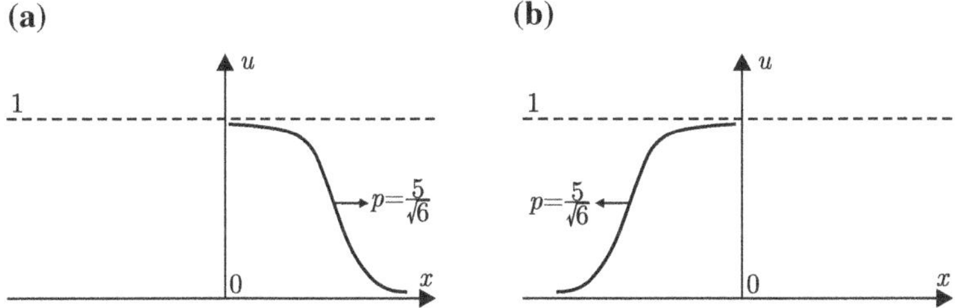

Fig. 3.16 Simple waves for Kolmogorov-Petrovsky-Piskunov-Fisher equation given by the explicit formula

Fig. 3.17 Approximated soliton-like solution to KPP-Fisher type equation, see Fig. 4.28 as well

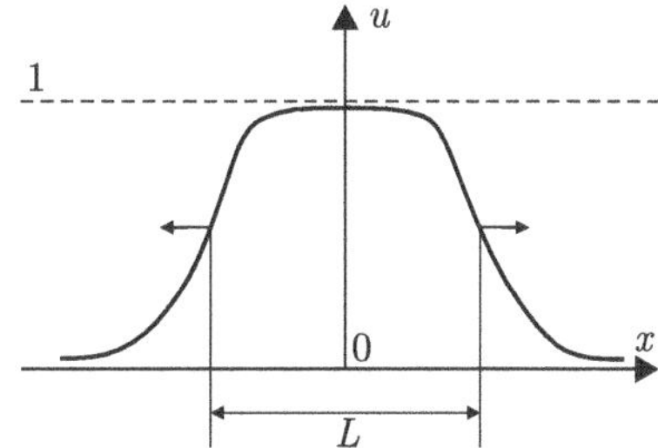

in the case where the distance between the wave fronts is large. Here there are several interesting effects which we consider for Eq. (3.169). For this equation, there exist two self-similar waves shown in Fig. 3.16. The product of these two functions is the function shown in Fig. 3.17. Obviously, this functions satisfies the equation up to $O(L^{-N})$, where $N > 0$ is an arbitrary number and L is the distance between the wave fronts.

In this case, the asymptotic law of wave superposition is a product, and we believe that an explicit expression for the leading term of the asymptotic solution can be obtained by multiplying explicit formulas for the solutions shown in Fig. 3.16 similarly to the method used for solitons, kinks, and the interaction of coalescence-type waves (in the last case, the asymptotic interaction law is the summation). But in the case of wave generation (decay), the leading term becomes

$$u = \left(1 + \chi^{-1/4}\psi(\sqrt{\chi_2})\exp\left(-\frac{\tau_1}{\sqrt{6}}\right)\right)^{-2}$$
$$\times \left(1 + \chi^{-1/4}\psi(\sqrt{\chi_1})\exp\left(-\frac{\tau_2}{\sqrt{6}}\right)\right)^{-2}$$
$$\times \left[1 + O(\exp\{-L/\sqrt{6}\})\right], \tag{3.172}$$

where $\chi_i = \chi(\tau_i)$, $i = 1, 2$, are self-similar waves whose graphs are shown in Fig. 3.16, $\chi(\tau) = \left(1 + \exp\left(-\tau/\sqrt{6}\right)\right)^{-2}$, the τ_i, $i = 1, 2$, are self-similar variables, for example, $\tau_1 = x + 5t/\sqrt{6}$, $\tau_2 = -x + 5/\sqrt{6} + L$, and $\psi(z)$ is a real solution of the problem

$$z^2(z-1)\psi'' + (9z-1)z\psi' - 8\psi = 0,$$
$$\psi(1) = 1, \qquad |\psi(0)| < \infty,$$

which can be expressed in terms of hypergeometric functions.

The asymptotic behavior of the function $\psi(z)$ is considered in [19, Sect. 2.5]. In particular, it follows from the conclusions in this section that, due to the initial conditions, the function u is close to the product in the domain between the wave fronts. Outside this domain (ahead of the wave fronts), the functions $\psi(\sqrt{\chi_i})$ oscillate, and ahead of the wave fronts, the function (3.172) can be represented in asymptotic form

$$u \cong U(\tau_1, \tau_2, i\tau_1, i\tau_2).$$

In Sect. 4.8, following the above reasoning, we consider the origination of the new (liquid) phase in the old (solid) phase and study the cathode melting. The nonlinearity of the equation in our model has the form shown in Fig. 3.12 for $A = -1$, $B = 0$, $C = 1$, and $F(\chi) = \chi(1 - \chi^2)$ in (3.170). In this case, as is shown in Fig. 3.14b, the perturbation concentrated in the strip $[-1, 0]$ attenuates (which means the stability of the constant solution $u = -1$) and develops for the values in the strip $[0, 1]$. If we consider the initial condition "mainly" concentrated in the strip $[-1, 0]$ but "a little" entering the strip $[0, 1]$, then it is an almost steady solution and its evolution is determined by external actions, i.e., the temperature in our case. For details, see Sect. 4.8.

References

1. Ablowitz, M.J., Zeppetella, A.: Explicit solutions of fisher's equation for a special wave speed. Bull. Math. Biol. **41**(6), 835–840 (1979)
2. Alexiades, V.: Mathematical Modeling of Melting and Freezing Processes. CRC Press (1992)
3. Alikakos, N.D., Bates, P.W.: On the singular limit in a phase field model of phase transitions. Annales de l'institut Henri Poincaré (C) Analyse non linéaire **5**(2), 141–178 (1988)
4. Bossavit, A., Damlamian, A., Fremond, M. (Eds.): Free Boundary Problems: Applications and Theory. Pitman (1985)
5. Caginalp, G.: Surface tension and supercooling in solidification theory. In: Garrido, L. (ed.) Applications of Field Theory to Statistical Mechanics. Lecture Notes in Physics, vol. 216, pp. 216–226. Springer, Berlin, Heidelberg (1985)
6. Caginalp, G.: An analysis of a phase field model of a free boundary. Arch. Ration. Mech. Anal. **92**(3), 205–245 (1986)
7. Caginalp, G.: The role of microscopic anisotropy in the macroscopic behavior of a phase boundary. Ann. Phys. **172**, 136–155 (1986)

8. Caginalp, G.: Stefan and Hele-shaw type models as asymptotic limits of the phase field equations. Phys. Rev. A **39**, 5887–5896 (1989)
9. Caginalp, G., Chadam, J.: Stability of interfaces with velocity correction term. Rocky Mount. J. Math. **21**(2), 617–629 (1991)
10. Caginalp, G., Chen, X.: Convergence of the phase field model to its sharp interface limits. Eur. J. Appl. Math. **9**(4), 417–445 (1998)
11. Caginalp, G., Fife, P.C.: Elliptic problems involving phase boundaries satisfying a curvature condition. IMA J. Appl. Math. **38**, 195–217 (1987)
12. Caginalp, G., McLeod, B.: The interior transition layer for ordinary differential equations arising from solidification theory. Quart. Appl. Math. **44**, 155–168 (1986)
13. Cariello, F., Tabor, M.: Painleve expansions for nonintegrable evolution equations. Phys. D: Nonlinear Phenom. **39**(1), 77–94 (1989)
14. Chadam, J., Howison, S.D., Ortoleva, P.: Existence and stability for spherical crystals growing in a supersaturated solution. IMA J. Appl. Math. **39**(1), 1–15 (1987)
15. Chalmers, B.: Principles of solidification. Wiley Series on the Science and Technology of Materials (Book 28). Wiley (1964)
16. Chen, X., Reitich, F.: Local existence and uniqueness of solutions of the stefan problem with surface tension and kinetic undercooling (November 1990). IMA Preprint Series 715
17. Crowley, A.B., Ockendon, J.R.: Modelling mushy regions. Appl. Sci. Res. **44**, 1–7 (1987)
18. Danilov, V.G.: On the relation between the Maslov-Whitham method and the weak asymptotics method. In: Kamiński, A., Oberguggenberger, M., Pilipović, S. (eds.) Linear and Non-Linear Theory of Generalized Functions and its Applications, vol. 88, pp. 55–65. Banach Center Publications, Warsaw (2010)
19. Danilov, V.G., Maslov, V.P., Volosov, K.A.: Mathematical Modelling of Heat and Mass Transfer Processes. Kluwer Academic Publication (1995)
20. Danilov, V.G., Omel'yanov, G.A., Radkevich, E.V.: Asymptotic behavior of the solution of a phase field system, and a modified stefan problem. Differ. Equat. **31**(3), 446–454 (1995)
21. Danilov, V.G., Omel'yanov, G.A., Radkevich, E.V.: Justification of asymptotics of solutions of the phase-field equations and a modified Stefan problem. Sbornik: Math. **186**(12), 1753–1771 (1995)
22. Danilov, V.G., Omel'yanov, G.A., Radkevich, E.V.: Hugoniot-type conditions and weak solutions to the phase-field system. Eur. J. Appl. Math. **10**, 55–77 (1999)
23. Danilov, V.G., Omel'yanov, G.A., Shelkovich, V.M.: Weak asymptotics method and interaction of nonlinear waves. Am. Math. Soc. Transl. 2, **208**, pp. 33–163. Providence: American Mathematical Society (2003)
24. Danilov, V.G., Subochev, P.Y.: Wave solutions of semilinear parabolic equations. Theor. Math. Phys. **89**, 1029–1046 (1991)
25. Egorov, Y.V.: Linear Differential Equations of Principal Type. Springer (1986)
26. Elliott, C.M., Ockendon, J.R.: Weak and Variational Methods for Free and Moving Boundary Problems. Pitman Publishing, Boston (1982)
27. Fife, P.C., Gill, G.S.: The phase-field description of mushy zones. Phys. D: Nonlinear Phenom. **35**, 267–275 (1989)
28. Fife, P.C., Gill, G.S.: Phase-transition mechanisms for the phase-field model under internal heating. Phys. Rev. A **43**(2), 843–851 (1991)
29. Gelfand, I.M., Shilov, G.E.: Generalized Functions: Properties and Operations. Academic Press (1964)
30. Gibbs, J.W.: The Collected Works. Yale University Press, New Haven (1948)
31. Glimm, J., Jaffe, A.: Quantum Physics: A Functional Integral Point of View, 2nd edn. Springer, NY (1987)
32. Hoffmann, K.H., Sprekels, J. (Eds.): Free Boundary Problems: Theory and Applications. Longman Scientific and Technical (1990)
33. Hohenberg, P.C., Halperin, B.I.: Theory of dynamic critical phenomena. Rev. Modern Phys. **49**(3), 435–479 (1977)

34. Howison, S.D., Lacey, A.A., Ockendon, J.R.: Hele-shaw free-boundary problems with suction. Quar. J. Mech. Appl. Math. **41**(2), 183–193 (1988)
35. Karma, A., Rappel, W.J.: Phase-field method for computationally efficient modeling of solidification with arbitrary interface kinetics. Phys. Rev. E **53**(4), R3017–R3020 (1996)
36. Kawahara, T., Tanaka, M.: Interactions of traveling fronts: An exact solution of a nonlinear diffusion equation. Phys. Lett. A **97**(8), 311–314 (1983)
37. Kolmogorov, A.N., Petrovskii, N.G., Piskunov, N.S.: A study of the diffusion equation with increase in the quantity of matter, and its application to a biological problem. Bull. Moscow State Univ. Ser. A. Math. Mech. **1**(6), 1–16 (1937). (in Russian)
38. Lacey, A.A., Tayler, A.B.: A mushy region in a Stefan problem. IMA J. Appl. Math. **30**(3), 303–313 (1983)
39. Lashin, A.M.: An investigation of the dynamics of first-order phase transition during the directional solidification of a pure metal into an undercooled melt on the base of phase-field model (2001). (in Russian)
40. Luckhaus, S.: Solutions for the two-phase stefan problem with the Gibbs–Thomson law for the melting temperature. Eur. J. Appl. Math. **1**(02), 101–111 (1990)
41. Luckhaus, S., Modica, L.: The Gibbs–Thompson relation within the gradient theory of phase transitions. Arch. Ration. Mech. Anal. **107**(1), 71–83 (1989)
42. Maslov, V.P., Omel'yanov, G.A.: Asymptotic soliton-form solutions of equations with small dispersion. Russian Math. Surv. **36**, 73–149 (1981)
43. Maslov, V.P., Tsupin, V.A.: Propagation of a shock wave in an isentropic gas with small viscosity. J. Soviet Math. **13**, 163–185 (1980)
44. Meirmanov, A.M.: An example of nonexistence of a classical solution of the Stefan problem. Soviet Math. Dokl. **23**, 564–566 (1981)
45. Modica, L.: The gradient theory of phase transitions and the minimal interface criterion. Arch. Ration. Mech. Anal. **98**(2), 123–142 (1987)
46. Oleinik, O.A.: Discontinuous solution of non-linear differential equations. AMS Transl. Ser. **2**(26), 95–172 (1963)
47. Oleinik, O.A., Radkevich, E.V.: On the analyticity of solutions of linear partial differential equations. Math. USSR–Sbornik **19**(4), 581–596 (1973)
48. Plotnikov, P.I., Starovoitov, V.N.: The Stefan problem with surface tension as a limit of the phase field model. Differ. Equat. **29**(3), 395–404 (1993)
49. Primicerio, M.: Mushy region in phase–change problem. Methoden und Verfahren der mathematischen Physik **25**, pp. 251–269. Peter Lang, Frankfurt/Main (1983)
50. Radkevich, E.V.: Gibbs–Thomson amendment and conditions for the existence of a classical solution of the modified stefan problem. Dokl. Akad. Nauk **316**(6), 1311–1315 (1991). (in Russian)
51. Radkevich, E.V.: About asymptotic solution of a phase-field system. Differ. Equat. **29**(3), 487–500 (1993)
52. Radkevich, E.V.: On conditions for the existence of a classical solution of the modified Stefan problem (the Gibbs–Thomson law). Russian Academy Sci. Sbornik Mathematics **75**, 221–246 (1993)
53. Radkevich, E.V.: On the heat Stefan wave. Dokl. Akad. Nauk USSR **47**(1), 150–155 (1993)
54. Soner, H.M.: Influence of the phase-field equations to the Mullins-Sekerka problem with kinetic undercooling. Arch. Ration. Mech. Anal. **131**(2), 139–197 (1995)
55. Treves, J.F.: Introduction to Pseudodifferential and Fourier Integral Operators, vol. 1: Pseudodifferential Operators, second printing edn. University Series in Mathematics. Plenum Press, NY (1982)
56. Uchiyama, K.: The behavior of solutions of some non-linear diffusion equations for large time. J. Math. Kyoto Univ. **18**(3), 453–508 (1978)
57. Visintin, A.: Models of Phase Transitions. Birkhäuser (1996)
58. Weiss, J., Tabor, M., Carnevale, G.: The painleve property for partial differential equations. J. Math. Phys. **24**(3), 528–552 (1983)

Chapter 4
Numerical Simulation and its Results

Abstract In the first part of this chapter, a numerical algorithm for solving the phase field system is presented with application to the real field emission nanocathode. The second part of this chapter contains the results of numerical simulations. In the third part of this chapter, we present an algorithm for introducing a liquid phase nucleus in the presented mathematical model of heat transfer in nanocathodes.

4.1 Nanocathode Model

As a spatial model of conic nanocathode (see Fig. 1.2; for the geometric dimensions, see Table 1.1 in Sect. 1.4), we consider the domain shown in Fig. 4.1. The spatial variables vary in the following ranges:

$$r \in [R_0, R], \quad \phi \in [0, \pi), \quad \vartheta \in [-\Theta, \Theta], \quad 0 < \Theta < \frac{\pi}{2}. \tag{4.1}$$

Figure 4.1 shows the free boundaries $r_1(t)$ and $r_2(t)$ which arise as the cathode melts, and Σ denotes the lateral surface of the cathode.

Thus, the domain where the model is defined is a truncated cone and its bases are given by the relations

$$r = R_0 \quad \text{and} \quad r = R.$$

Such a choice of the modeling domain permits approximating the conic nanocathode in the simplest way using the spherical coordinates.

All values of the physical parameter used in experiments [5] which we use in numerical simulation are given in Table 4.1.

The process of nanocathode heating by the electric current passing through it is described by heat equation (1.2).

The temperature is made dimensionless by the formula

© Springer Nature Singapore Pte Ltd. 2020

V. Danilov et al., *Mathematical Modeling of Emission in Small-Size Cathode*,
Heat and Mass Transfer, https://doi.org/10.1007/978-981-15-0195-1_4

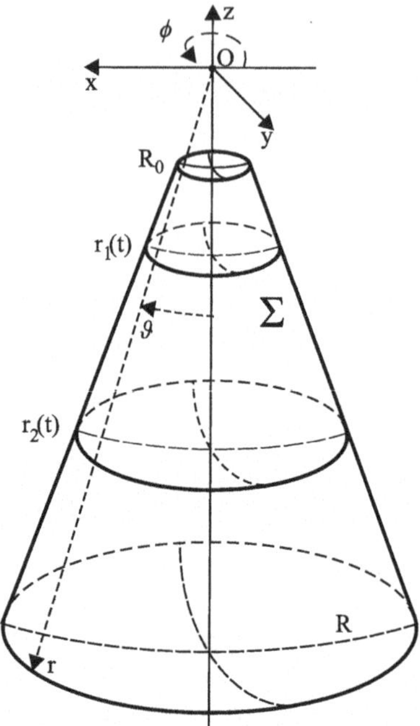

Fig. 4.1 Nanocathode model

Table 4.1 Physical parameters of the model according to [5]

t_0	Characteristic time (time of experiment)	100 s
r_0	Characteristic scale (cathode size)	10^{-5} m
l	Latent heat of melting	1.64×10^5 J/kg
λ	Specific thermal conductivity	149 W/(m $\cdot$ K)
c	Specific heat capacity	678 J/(kg $\cdot$ K)
e	Electron charge (without sign)	1.602×10^{-19} C
T_0	Melting temperature	1683 K
σ_{SB}	Stefan–Boltzmann constant	5.67×10^{-8} J/(m^{2}K^{4}s)
ς	Electron work function	4.8 eV
k_B	Boltzmann constant	1.381×10^{-23} J/K
σ	Surface tension	0.725 N/m
ρ	Cathode material density	2330 kg/m^3
μ	Kinetic growth coefficient	0.5 m/(s $\cdot$ K)
k	Thermal conductivity ($k = \lambda/(c\rho)$)	9.43×10^{-5} m^2/s

$$\bar{\theta} = \frac{c}{l}T, \qquad (4.2)$$

where $\bar{\theta}$ is the dimensionless temperature. The dimensionless time t and the coordinate (radius) r are determined by the expressions

$$\tilde{t} = t_0 t, \qquad \tilde{r} = r_0 r, \qquad (4.3)$$

where $\tilde{t}$ is the dimensional time (measured in seconds) and $\tilde{r}$ is the dimensional radius (measured in meters). With regard to (1.4), (4.2), and (4.3), we derive from Eq. (1.2) the dimensionless heat conduction equation in spherical coordinates

$$\frac{\partial \bar{\theta}}{\partial t} - \hat{k}\Delta\bar{\theta} = \hat{F}, \qquad (4.4)$$

where the Laplace operator has the form

$$\Delta = \frac{1}{r^2}\frac{\partial}{\partial r}\left(r^2\frac{\partial}{\partial r}\right) + \frac{1}{r^2\sin\vartheta}\frac{\partial}{\partial\vartheta}\left(\sin\vartheta\frac{\partial}{\partial\vartheta}\right) + \frac{1}{r^2\sin^2\vartheta}\frac{\partial^2}{\partial\phi^2}, \qquad (4.5)$$

$$\hat{k} = \frac{\lambda}{\rho c}\frac{t_0}{r_0^2} = k\frac{t_0}{r_0^2}, \qquad \hat{F} = \frac{t_0}{l\rho}\frac{j_{\mathrm{in}}^2}{\sigma_e}, \qquad (4.6)$$

the specific conductivity σ_e is determined by formula (2.24) (see Fig. 4.31, where the jump of conductivity after the cathode melting is shown), and the current density inside the cathode j_{in} is determined from Eq. (1.3) (see Sect. 4.2 below for details, the physical parameters are given in Table 4.1).

A specific feature of Eq. (4.4) is that the dimensionless thermal conductivity is large $\left(\hat{k} \approx 10^8\right)$. This means that the temperature levels very fast in a small volume and enters a quasistationary solution.

The dimensionless Stefan condition (1.6) in spherical coordinates becomes

$$k\left[\frac{\partial\bar{\theta}}{\partial r}\right]\Bigg|_{r=r_i(t)} = (-1)^{i+1}\dot{r}_i(t)r_i(t), \qquad (4.7)$$

where $r = r_i(t)$, $i = 1, 2$, are free boundaries, see Fig. 4.1.

The dimensionless Gibbs–Thomson condition (1.7) becomes

$$\left(\bar{\theta} - \bar{\theta}_0\right)\Bigg|_{r=r_i(t)} = (-1)^i\left(\frac{\hat{\alpha}}{\hat{\beta}}\dot{r}_i(t) + \hat{\beta}\frac{2}{r_i(t)}\right), \qquad i = 1, 2, \qquad (4.8)$$

where

$$\bar{\theta}_0 = \frac{c}{l}T_0, \qquad \hat{\alpha} = \frac{\hat{\beta}}{\mu(l/c)}\frac{r_0}{t_0}, \qquad \hat{\beta} = \frac{\sigma\bar{\theta}_0}{l\rho}\frac{1}{r_0}; \qquad (4.9)$$

here $\bar{\theta}_0$ is the dimensionless melting temperature.

In our case, the coefficients in condition (4.8) take the following values: $\hat{\alpha} = 10^{-12}$ and $\hat{\beta} \approx 10^{-3}$. Thus, we obtain the relation

$$\hat{\alpha} \ll \hat{\beta}. \tag{4.10}$$

Relation (4.10) means that, on the upper and lower free boundaries (see Fig. 4.1), the Gibbs–Thomson conditions (4.8) actually have the form

$$\left.(\bar{\theta} - \bar{\theta}_0)\right|_{r=r_1(t)} = -\hat{\beta}\frac{2}{r_1(t)}, \quad \left.(\bar{\theta} - \bar{\theta}_0)\right|_{r=r_2(t)} = \hat{\beta}\frac{2}{r_2(t)}. \tag{4.11}$$

This means that the influence of the curvature on the free boundary is much greater than the influence of the velocity.

As a regularization of problem (4.4), (4.7), (4.11) with a small parameter ε, we use the phase field system (3.13). In our case, it becomes

$$\frac{\partial \theta}{\partial t} - \hat{k}\Delta\theta = -\frac{1}{2}\frac{\partial \varphi}{\partial t} + \hat{F}, \tag{4.12}$$

$$\varepsilon\hat{\alpha}\frac{\partial \varphi}{\partial t} - \varepsilon\hat{\beta}^2\Delta\varphi = \frac{1}{\varepsilon}\left(\varphi - \varphi^3\right) + \chi\left(1 - \varphi^2\right)\left(\theta - \bar{\theta}_0\right). \tag{4.13}$$

Here the function $\theta = \theta(r, \phi, \vartheta, t, \varepsilon)$ is a regularization of the temperature $\bar{\theta} = \bar{\theta}(r, \phi, \vartheta, t)$, $\varphi = \varphi(r, \phi, \vartheta, t, \varepsilon)$ is the function of "order". We shall explain the last relation. From the definition of generalized solution (see Sect. 3.4.1, Definition 3.2), for Eq. (4.13) we obtain the Stefan–Gibbs–Thomson condition on the free boundaries

$$\left.a_2(\theta - \bar{\theta}_0)\right|_{r=r_i(t)} = (-1)^i\left(\frac{\hat{\alpha}}{\hat{\beta}}\dot{r}_i(t) + \hat{\beta}\frac{2}{r_i(t)}\right)a_1, \quad i = 1, 2,$$

where

$$a_1 = \int \dot{w}^2(\tau)d\tau, \quad a_2 = \chi\int \dot{w}(\tau)(1 - w^2(\tau))d\tau, \quad w(\tau) = \tanh(\tau).$$

Since $\hat{\alpha}$ is small, to obtain a Stefan-type condition in the form (4.11), we must put

$$\chi = \frac{\int \dot{w}^2(\tau)d\tau}{\int \dot{w}(\tau)(1 - w^2(\tau))d\tau}.$$

We calculate the integral to obtain $\chi = 1$.

As was already noted, the solution of Eq. (4.12) rapidly becomes stationary. With regard to (4.10), we see that the solution of Eq. (4.13) also becomes stationary at a prescribed temperature.

Equation (4.12) is supplemented with the following boundary conditions. On the upper base (i.e., at the top of the cathode), we pose a boundary condition that determines the Nottingham effect (as was noted in Sect. 2.4, this is the cooling condition) whose explicit form is given further in Sect. 4.3 [see (4.34)]. On the lower base, we pose either the cooling condition

$$\left. \frac{\partial \theta}{\partial r} \right|_{r=R} = -\alpha_{\text{cool}}, \tag{4.14}$$

where $\alpha_{\text{cool}} > 0$, or the heat insolation condition which is a particular case of condition (4.14) for $\alpha_{\text{cool}} = 0$, i.e., the Neumann condition. The cooling through the lower cathode base is possible through the heat withdrawal from the substratum on which the cathode is located (see Fig. 1.2) but the value of this possible cooling is small. On the cathode lateral surface, we pose the heat insolation condition, i.e., the Neumann condition

$$\left. \frac{\partial \theta}{\partial \mathbf{n}} \right|_{\Sigma} = \left. \frac{\partial \theta}{\partial \vartheta} \right|_{\vartheta=\pm\Theta} = 0. \tag{4.15}$$

Equation (4.13) is supplemented with Neumann boundary conditions, see Sect. 4.4 for details.

4.2 Calculation of the Current Density Inside the Cathode

As was previously noted, the electric current inside the cathode is described by Eq. (1.3):

$$\text{div } \mathbf{j}_{\text{in}} = 0, \tag{4.16}$$

where the current density inside the cathode $\mathbf{j}_{\text{in}}$ is expressed through the potential by formula (1.5):

$$\mathbf{j}_{\text{in}} = -\sigma_{\text{e}}(T)\left[\nabla\Psi + A(T)\nabla T\right];$$

here T is the real temperature (in Kelvins). The specific conductivity $\sigma_{\text{e}}(T)$ is described by formula (2.24), and the thermal EMF coefficient $A(T)$ is given by formula (2.29). As was previously noted and shown in Sect. 2.3, the coefficient $A(T)$ is small for silicon and can be neglected, i.e.,

$$\mathbf{j}_{\text{in}} = -\sigma_{\text{e}}(T)\nabla\Psi. \tag{4.17}$$

Thus, we obtain the following equation for the potential Ψ:

$$\langle \nabla, \sigma_e(T)\nabla\Psi \rangle = 0. \tag{4.18}$$

Taking into account that the cathode is axially symmetric (i.e., the domain where the modeling is carried out is independent of the polar angle ϕ) and passing to spherical coordinates in Eq. (4.18), in which the Laplace operator is described by formula (4.5) and the nabla operator has the form

$$\nabla = \left(\frac{\partial}{\partial r}, \frac{1}{r}\frac{\partial}{\partial \vartheta}, \frac{1}{r\sin\vartheta}\frac{\partial}{\partial \phi} \right),$$

we obtain the following two-dimensional equation for the potential:

$$(L_1 + L_2)\Psi = 0, \tag{4.19}$$

where

$$L_1 = \frac{1}{r^2}\frac{\partial}{\partial r}\left(\sigma_{\mathrm{e}} r^2 \frac{\partial}{\partial r} \right); \quad L_2 = \frac{1}{r^2 \sin\vartheta}\frac{\partial}{\partial \vartheta}\left(\sigma_{\mathrm{e}} \sin\vartheta \frac{\partial}{\partial \vartheta} \right).$$

The boundary conditions for Eq. (4.19) have the form

$$\Psi \Big|_{r=R_0} = \Psi_1 = 0, \tag{4.20}$$

$$\Psi \Big|_{r=R} = \Psi_2, \tag{4.21}$$

$$\frac{\partial \Psi}{\partial \vartheta} \Big|_{\vartheta=\pm\Theta} = 0. \tag{4.22}$$

Condition (4.21) means that the potential Ψ_2 is applied to the cathode lower base, and condition (4.20) means that the potential at the cathode spot is equal to $\Psi_1 = 0\,\mathrm{V}$ (in fact, it is zero at the anode, but since the distance between anode and cathode is small, we can neglect the latter in the problem under study, see Sect. 4.6). On the cathode lateral surface, we pose the Neumann condition (4.22).

Equation (4.19) is stationary. We shall solve it numerically by the iteration method [11]. For this, we introduce a new function $\hat{\Psi} = \hat{\Psi}(r, \vartheta, \check{t})$ satisfying the nonstationary equation

$$\frac{\partial \hat{\Psi}}{\partial \check{t}} = (L_1 + L_2)\hat{\Psi}, \tag{4.23}$$

where $\check{t}$ is the "iteration" time. Equation (4.23) is supplemented with similar (4.20)–(4.22) boundary conditions for the function $\hat{\Psi}$:

$$\hat{\Psi}\Big|_{r=R_0} = \Psi_1 = 0, \tag{4.24}$$

$$\hat{\Psi}\Big|_{r=R} = \Psi_2, \tag{4.25}$$

$$\frac{\partial \hat{\Psi}}{\partial \vartheta}\Big|_{\vartheta=\pm\Theta} = 0. \tag{4.26}$$

As the initial condition for Eq. (4.23) we take

$$\hat{\Psi}(r, \vartheta, 0) = 0. \tag{4.27}$$

We assume that

$$\Psi(r, \vartheta) = \hat{\Psi}(r, \vartheta, t^*).$$

where t^* is determined from the condition that the solution becomes stationary:

$$\left|\hat{\Psi}(r, \vartheta, t^*) - \hat{\Psi}(r, \vartheta, \check{t})\right| < \tilde{\varepsilon} \quad \text{for } \forall \check{t} > t^*; \tag{4.28}$$

here $\tilde{\varepsilon}$ is a prescribed accuracy. The difference scheme for numerically solving problem (4.23), (4.24)–(4.27) is given in Sect. 4.4.1.

If the potential is known at each point of the cathode, then we can easily determine the current density inside the cathode. It is well known that

$$\mathbf{E}_F = -\nabla\Psi, \tag{4.29}$$

where $\mathbf{E}_F$ is the electric field strength. Taking the last formula and formula (4.17) into account, we obtain

$$\mathbf{j}_{\text{in}} = -\sigma_{\text{e}}(T)\mathbf{E}_F. \tag{4.30}$$

Now we can calculate the heat release power density F:

$$F = |\mathbf{j}_{\text{in}}|^2/\sigma_{\text{e}}(T). \tag{4.31}$$

Thus, if the real temperature T related to the dimensionless θ by formula (4.2) is known at the current time step, then calculating the values of the potential in Eq. (4.23) on the spatial grid and using the standard numerical methods, we can calculate the strength of the electric field by formula (4.29), the current density by formula (4.30), and the heat release power density by formula (4.31), see Sect. 4.4 for details.

4.3 Calculation of the Emission Current Density and Modeling of the Nottingham Effect

We consider boundary condition (1.10) which determines the heat balance at the top of the cathode:

$$\lambda \frac{\partial T}{\partial \tilde{r}}\bigg|_{\tilde{r}=\tilde{R}_0} = \frac{j_{\text{em}}}{e}\mathcal{E}_{\text{Nott}}\bigg|_{\tilde{r}=\tilde{R}_0}. \tag{4.32}$$

Here, as was already noted, j_{em} is the emission current density, $\mathcal{E}_{\text{Nott}}$ is the average energy of emitting electrons, T is the real temperature (in Kelvins), $\tilde{r}$ is the real coordinate (in meters), and the other parameters are given in Table 4.1.

First, we determine the emission current density j_{em} by formula (2.104). To this end, we calculate the integral for different values of the temperature and the strength of electric field approximately by using the Simpson method [7]. Further, in the algorithm for numerical modeling (see Sects. 4.4 and 4.5), we use the table of the integral values calculated in advance for different temperature and field strength, and to determine the integral values, we bilinearly interpolate the values in the table.

Now it is necessary to calculate the average energy of emitting electrons $\mathcal{E}_{Nott}$ in formula (4.32). To calculate the energy, we use approximate formula (2.117) in dimensionless form (see above in Sect. 2.4):

$$\mathcal{E} = \begin{cases} -0.0589529\,\theta\cot(14.6137\,\theta), & \theta \leqslant \theta^*, \\[2mm] 4.42871 + 0.0417038\,\theta \\[1mm] \quad + \dfrac{-21.8518 + 0.25058\,\theta}{4.92306 + (-1 + 9.30338\,\theta)^{3.48481}}, & \theta > \theta^*. \end{cases} \tag{4.33}$$

Here $\mathcal{E}$ is the dimensionless form of $\mathcal{E}_{Nott}$, and the dimensionless inversion temperature θ^* is related to the real inversion temperature T^* by the formula

$$\theta^* = \frac{c}{l}T^*.$$

In the case under study, the dependence of T^* on the strength of external field is illustrated in Fig. 2.18 in Sect. 2.4.9, which means that the Nottingham effect is the cooling at the cathode temperature near the melting temperature, see Sect. 2.4.

As a result, boundary condition (4.32) can be written in dimensionless form as

$$\left.\frac{\partial\theta}{\partial r}\right|_{r=R_0} = \frac{c}{l}\frac{r_0}{\lambda}\frac{j_{em}}{e}\mathcal{E}\Big|_{r=R_0}. \tag{4.34}$$

4.4   Difference Scheme

We note that the domain where the modeling is carried out (see (4.1) and Fig. 4.1 in Sect. 4.1) is axially symmetric (i.e., independent of the angle $\phi \in [0, 2\pi)$). It follows from this fact and the form of the boundary conditions that the whole problem is axially symmetric. A partial melting of the cathode is possible when the current passes through it. Clearly, by the symmetry, the domains of the same phase form layers inside the cathode.

If the heat can be removed (for example, due to radiation) from the lateral surface of the cone, then the liquid phase is first formed inside the cathode with subsequent (possible) exit of the free surface to the lateral surface. Of course, the boundary condition on the order function

$$\left.\frac{\partial \varphi}{\partial \mathbf{n}}\right|_{\Gamma} = 0 \tag{4.35}$$

(where $\Gamma = \Sigma \cup \{r = R_0\} \cup \{r = R\}$) cannot be satisfied at the time and point of contact between the free boundary and the boundary of the domain. But if we neglect the gravity force, then we can still assume that, in the case of further melting, the melting region is fast transformed in a layer $0 < R_0 \leqslant r \leqslant R$ such that condition (4.35) is satisfied on its boundaries.

We start to construct the difference scheme for solving phase field system (4.12), (4.13) which, for convenience, we write as

$$\frac{\partial \theta}{\partial t} - \hat{k}(A_1 + A_2)\theta = -\frac{1}{2}\frac{\partial \varphi}{\partial t} + \hat{F}, \tag{4.36}$$

$$\varepsilon \hat{\alpha}\frac{\partial \varphi}{\partial t} - \varepsilon \hat{\beta}^2(A_1 + A_2)\varphi = \frac{1}{\varepsilon}g(\varphi) + \chi f(\varphi)(\theta - \bar{\theta}_0), \tag{4.37}$$

where

$$A_1 = \frac{1}{r^2}\frac{\partial}{\partial r}\left(r^2\frac{\partial}{\partial r}\right); \quad A_2 = \frac{1}{r^2 \sin \vartheta}\frac{\partial}{\partial \vartheta}\left(\sin \vartheta \frac{\partial}{\partial \vartheta}\right),$$

$$g(\varphi) = \varphi - \varphi^3; \quad f(\varphi) = 1 - \varphi^2. \tag{4.38}$$

Further, in this section, we use the well-known methods for constructing difference schemes for solving systems of equations numerically by the alternating direction implicit (ADI) method [11]. First, we make several remarks.

Remark 4.1 The system of Eqs. (4.36), (4.37) is two-dimensional in the spatial variables. Therefore, we use the following well-known fact to construct the difference scheme. Assume that we have a two-dimensional equation with right-hand side

$$D_t(w) + D(w) = F(w), \tag{4.39}$$

where D_t is the differential operator with respect to the time variable, $D = D_{xx} + D_{yy}$ is the differential operator representing a sum of differential operators with respect to distinct spatial variables, and ω is the desired function. By the ADI method, Eq. (4.39) can be associated with following difference equations [11]:

$$\begin{cases} \bar{D}_t(w^{k+1/2}, w^k) + \bar{D}_{xx}(w^{k+1/2}) = F(w^{k+1/2}), \\ \bar{D}_t(w^{k+1}, w^{k+1/2}) + \bar{D}_{yy}(w^{k+1}) = 0, \end{cases} \tag{4.40}$$

where $\bar{D}_t$, $\bar{D}_{xx}$, and $\bar{D}_{yy}$ are the corresponding difference operators and ω^k is the desired functions at the (k)th time layer. The second version is possible:

$$\begin{cases} \bar{D}_t(w^{k+1/2}, w^k) + \bar{D}_{xx}(w^{k+1/2}) = \frac{1}{2}F(w^{k+1/2}), \\ \bar{D}_t(w^{k+1}, w^{k+1/2}) + \bar{D}_{yy}(w^{k+1}) = \frac{1}{2}F(w^k). \end{cases}$$

To solve this numerical problem, we use the first version, i.e., the system of difference equations (4.40).

Remark 4.2 A certain differential operator

$$D_{xx} = \frac{\partial}{\partial x}\left(q(x)\frac{\partial p}{\partial x}\right)$$

is approximated by the expression

$$\bar{D}_{xx} = \frac{1}{h}\left(a_{i+1}\frac{p_{i+1} - p_i}{h} - a_i\frac{p_i - p_{i-1}}{h}\right),$$

where the coefficients a_i can be determined in different ways [11], for example, as

$$a_i = (q_{i-1} + q_i)/2 \tag{4.41}$$

or

$$a_i = q_{(i-1)/2}. \tag{4.42}$$

The method chosen to calculate a_i significantly influences the accuracy of computations. Formula (4.41) is better than (4.42) (with respect to the accuracy) [11], and hence we use it.

Remark 4.3 System (4.36), (4.37) is nonlinear. The nonlinearity is contained in the second equation of this system, namely, in the functions (4.38). We realize these nonlinearities by the Taylor formula:

$$g(\varphi_{i,j}^{k+1}) = g(\varphi_{i,j}^k) + g'(\varphi_{i,j}^k)\big(\varphi_{i,j}^{k+1} - \varphi_{i,j}^k\big), \tag{4.43}$$

$$f(\varphi_{i,j}^{k+1}) = f(\varphi_{i,j}^k) + f'(\varphi_{i,j}^k)\big(\varphi_{i,j}^{k+1} - \varphi_{i,j}^k\big). \tag{4.44}$$

Remark 4.4 System (4.36), (4.37) is coupled. To split the system, we first solve difference equation (4.37) for a given grid temperature at the preceding step. After the value of the difference order function is determined at the subsequent time step from Eq. (4.37), we obtain the value of the grid temperature at the subsequent time from the difference heat equation (4.36).

The approach described above gives a rather good result in numerical modeling of problems with a free boundary for the phase field system (see, e.g., [9]).

In our domain of modeling

$$r \in [R_0, R]; \quad R_0 > 0; \quad \vartheta \in [-\Theta, \Theta], \; 0 < \Theta < \pi/2; \quad t \in [0, t_{\max}],$$

we introduce the uniform grid ω with respect to the coordinates r, ϑ and the time t:

$$\omega = \omega_r \times \omega_\vartheta \times \omega_t :$$

$$\omega_r = \left\{ r_i = R_0 + ih_r, \ h_r > 0, \ i = 0, \dots, I \right\}; \tag{4.45}$$

$$\omega_\vartheta = \left\{ \vartheta_j = -\Theta + jh_\vartheta, \ h_\vartheta > 0, \ j = 0, \dots, J \right\}, \tag{4.46}$$

$$\omega_t = \left\{ t^k = kh_t, \ h_t > 0, \ k = 0, \dots, K \right\}, \tag{4.47}$$

where

$$I = \left\lfloor \frac{R - R_0}{h_r} \right\rfloor, \ J = \left\lfloor \frac{2\Theta}{h_\vartheta} \right\rfloor, \ K = \left\lfloor \frac{t_{\max}}{h_t} \right\rfloor.$$

Here h_r is the step of the spatial grid in the coordinate r, h_ϑ is the step of the spatial grid in the angle ϑ, h_t is the grid step with respect to time, and $t_{\max}$ is the whole time of the problem modeling. Here we also use the notation $\lfloor x \rfloor = \max\{n \in \mathbb{Z} \mid n \leqslant x\}$ to denote the greatest integer x less than or equal to x. We introduce the grid functions $\varphi_{i,j}^k = \varphi(r_i, \vartheta_j, t^k)$, $\theta_{i,j}^k = \theta(r_i, \vartheta_j, t^k)$, $\Psi_{i,j}^k = \Psi(r_i, \vartheta_j, t^k)$, $\sigma_{i,j}^k = \sigma_{\mathrm{e}}\big((l/c)\theta_{i,j}^k\big)$.

Now we shall construct difference approximations for Eqs. (4.36), (4.37), (4.23) with corresponding boundary and initial conditions. Further, we use the difference schemes given below to describe the algorithm for numerically solving the complete mathematical model for the problem under study.

4.4.1 Difference Scheme for the Equation for the Potential

We construct the difference scheme for the initial boundary value problem for the equation for the potential (4.23), (4.24)–(4.27) according to the above remarks 4.1–4.4. We introduce a uniform grid in the "iteration" time $\check{t}$ with step τ:

$$\omega_\tau = \left\{ \check{t}^n = \tau n, \ \tau > 0, \ n = 0, 1, 2, \dots \right\}.$$

We introduce the grid function $\hat{\Psi}_{i,k}^n = \hat{\Psi}(r_i, \vartheta_j, \check{t}^n)$. The initial condition (4.27) $\hat{\Psi}(r, \vartheta, 0) = 0$ becomes

$$\hat{\Psi}_{i,j}^0 = 0. \tag{4.48}$$

The equation for the potential (4.23) $\partial \hat{\Psi} / \partial \check{t} + (L_1 + L_2)\hat{\Psi} = 0$, where

$$L_1 = \frac{1}{r^2} \frac{\partial}{\partial r}\left(\sigma_{\mathrm{e}} r^2 \frac{\partial}{\partial r} \right), \quad L_2 = \frac{1}{r^2 \sin \vartheta} \frac{\partial}{\partial \vartheta}\left(\sigma_{\mathrm{e}} \sin \vartheta \frac{\partial}{\partial \vartheta} \right),$$

is approximated by the following system of difference equations.

The first equation in this system has the form

$$\frac{\hat{\Psi}_{i,j}^{n+1/2} - \hat{\Psi}_{i,j}^{n}}{\tau} = \frac{1}{r_i^2} \frac{1}{2h_r^2}\left[\left(\sigma_{i+1,j}^k r_{i+1}^2 + \sigma_{i,j}^k r_i^2\right)\left(\hat{\Psi}_{i+1,j}^{n+1/2} - \hat{\Psi}_{i,j}^{n+1/2}\right)\right.$$
$$\left. - \left(\sigma_{i,j}^k r_i^2 + \sigma_{i-1,j}^k r_{i-1}^2\right)\left(\hat{\Psi}_{i,j}^{n+1/2} - \hat{\Psi}_{i-1,j}^{n+1/2}\right)\right], \tag{4.49}$$

where $i = 1, 2, \ldots, I-1$, $j = 0, 1, \ldots, J$, $n > 0$, and the second equation has the form

$$\frac{\hat{\Psi}_{i,j}^{n+1} - \hat{\Psi}_{i,j}^{n+1/2}}{\tau} = \frac{1}{r_i^2 \sin\vartheta_j} \frac{1}{2h_\vartheta^2}\left[\left(\sigma_{i,j+1}^k \sin\vartheta_{j+1} + \sigma_{i,j}^k \sin\vartheta_j\right)\left(\hat{\Psi}_{i,j+1}^{n+1} - \hat{\Psi}_{i,j}^{n+1}\right)\right.$$
$$\left. - \left(\sigma_{i,j}^k \sin\vartheta_j + \sigma_{i,j-1}^k \sin\vartheta_{j-1}\right)\left(\hat{\Psi}_{i,j}^{n+1} - \hat{\Psi}_{i,j-1}^{n+1/2}\right)\right], \tag{4.50}$$

where $i = 0, 1, \ldots, I$, $j = 1, 2, \ldots, J-1$, $n > 0$.

Boundary conditions (4.24)–(4.26)

$$\hat{\Psi}\big|_{r=R_0} = \Psi_1 = 0, \quad \hat{\Psi}\big|_{r=R} = \Psi_2 = \text{const}, \quad \left(\partial\hat{\Psi}/\partial\vartheta\right)\big|_{\vartheta=\pm\Theta} = 0$$

are approximated by the difference expressions

$$\hat{\Psi}_{0,j}^{n+1/2} = \Psi_1 = 0, \tag{4.51}$$
$$\hat{\Psi}_{I,j}^{n+1/2} = \Psi_2 = \text{const}, \tag{4.52}$$
$$\hat{\Psi}_{i,0}^{n+1} = \hat{\Psi}_{i,1}^{n+1}, \quad \hat{\Psi}_{i,J}^{n+1} = \hat{\Psi}_{i,J-1}^{n+1}. \tag{4.53}$$

The calculations by the difference scheme (4.48)–(4.53) are carried out until the solution becomes stationary. The stationarity criterion (4.28) in difference form becomes

$$\left|\hat{\Psi}_{i,j}^{n^*} - \hat{\Psi}_{i,j}^{m}\right| < \tilde{\varepsilon} \quad \text{for } \forall\, m > n^*,$$

where $\tilde{\varepsilon}$ is a prescribed accuracy. As a result, we obtain the potential at the kth step with respect to the "global" time

$$\Psi_{i,j}^k = \hat{\Psi}_{i,j}^{n^*}. \tag{4.54}$$

Now it is easy to determine the electric field $\mathbf{E}_F$ by formula (4.29) which in difference form becomes

$$\left(\mathbf{E}_F\right)_{i,j}^k = -\left(\tilde{E}_{i,j}^k, \frac{1}{r_i}\hat{E}_{i,j}^k\right). \tag{4.55}$$

Here the components $\tilde{E}_{i,j}^k$ and $\hat{E}_{i,j}^k$ at the inner points of the difference grid have the form

$$\tilde{E}_{i,j}^{k} = \frac{\Psi_{i+1,j}^{k} - \Psi_{i-1,j}^{k}}{2h_r}, \quad \hat{E}_{i,j}^{k} = \frac{\Psi_{i,j+1}^{k} - \Psi_{i,j-1}^{k}}{2h_\vartheta},$$

where $i = 1, 2, \ldots, I - 1$, $j = 1, 2, \ldots, J - 1$, and at the boundary points, they have the form

$$\tilde{E}_{0,j}^{k} = \frac{\Psi_{1,j}^{k} - \Psi_{0,j}^{k}}{h_r}, \quad \tilde{E}_{I,j}^{k} = \frac{\Psi_{I,j}^{k} - \Psi_{I-1,j}^{k}}{h_r}, \quad j = 0, 1, \ldots J,$$

$$\hat{E}_{i,0}^{k} = \frac{\Psi_{i,1}^{k} - \Psi_{i,0}^{k}}{h_\vartheta}, \quad \hat{E}_{i,J}^{k} = \frac{\Psi_{i,J}^{k} - \Psi_{i,J-1}^{k}}{h_\vartheta}, \quad i = 0, 1, \ldots, I.$$

Then it is easy to obtain the current density by formula (4.30) which in difference form becomes

$$\left(\mathbf{j}_{\text{in}}\right)_{i,j}^{k} = -\sigma_{i,j}^{k}\left(\mathbf{E}_F\right)_{i,j}^{k}, \quad i = 0, 1, \ldots I, \; j = 0, 1, \ldots, J. \tag{4.56}$$

As a result, with regard to (4.31) we obtain the following formula for calculating the heat release power density F:

$$F_{i,j}^{k} = \sigma_{i,j}^{k}\left(\left(\tilde{E}_{i,j}^{k}\right)^2 + \left(\frac{1}{r_i}\hat{E}_{i,j}^{k}\right)^2\right), \tag{4.57}$$

where $i = 0, 1, \ldots I$, $j = 0, 1, \ldots, J$.

4.4.2 Difference Scheme for the Equation for the Order Function

The order function is determined from Eq. (4.37)

$$\varepsilon\hat{\alpha}\frac{\partial\varphi}{\partial t} - \varepsilon\hat{\beta}^2(A_1 + A_2)\varphi = \frac{1}{\varepsilon}g(\varphi) + \chi f(\varphi)(\theta - \bar{\theta}_0),$$

supplemented with Neumann boundary conditions (4.35). As the initial condition, one usually takes the solid phase, i.e.,

$$\varphi\big|_{t=0} = -1.$$

We note that the coefficient at $\partial\varphi/\partial t$ in Eq. (4.37) is small (the coefficient $\hat{\alpha}$ is small, see (4.10), and ε is a small number). Therefore, the solution of Eq. (4.37) is stationary, and hence, instead of Eq. (4.37) we shall solve the stationary equation

$$-\varepsilon\hat{\beta}^2(A_1 + A_2)\varphi = \frac{1}{\varepsilon}g(\varphi) + \chi f(\varphi)(\theta - \bar{\theta}_0), \tag{4.58}$$

where the operators A_1 and A_2 are defined in the preceding subsection. The method used to solve equations of such a type is described in detail for an example of equation for the potential in Sect. 4.2 and in Sect. 4.4.1.

Similarly, instead of stationary equation (4.58), we consider the nonstationary equation

$$\frac{\partial \hat{\varphi}}{\partial \check{t}} - \varepsilon \hat{\beta}^2 (A_1 + A_2) \hat{\varphi} = \frac{1}{\varepsilon} g(\hat{\varphi}) + \chi f(\hat{\varphi})(\theta - \bar{\theta}_0).$$

Remark 4.5 Formally, this means that we solve original problem (4.37) for $\varepsilon \alpha = 1$ till the moment of stabilization.

Following a similar scheme, as in Sect. 4.4.1, we introduce a uniform grid ω_τ with respect to the "iteration" time $\check{t}$. We introduce the grid function $\hat{\varphi}_{i,j}^n = \hat{\varphi}(r_i, \vartheta_j, \check{t}^n)$. The initial conditions have the form

$$\hat{\varphi}_{i,j}^0 = \varphi_{i,j}^k. \tag{4.59}$$

(Here φ^k is the function at the kth "global" time step).

Equation (4.58) is approximated by the following system of difference equations. The first equation in this system has the form

$$\frac{\hat{\varphi}_{i,j}^{n+1/2} - \hat{\varphi}_{i,j}^n}{\tau} - \varepsilon \hat{\beta}^2 \frac{1}{r_i^2} \frac{1}{2h_r^2} \Big[(r_{i+1}^2 + r_i^2)(\hat{\varphi}_{i+1,j}^{n+1/2} - \hat{\varphi}_{i,j}^{n+1/2})$$

$$- (r_i^2 + r_{i-1}^2)(\hat{\varphi}_{i,j}^{n+1/2} - \hat{\varphi}_{i-1,j}^{n+1/2}) \Big] = \frac{\hat{\varphi}_{i,j}^{n+1/2}}{\varepsilon} \big(1 - 3(\hat{\varphi}_{i,j}^n)^2 \big) + \frac{2}{\varepsilon} (\hat{\varphi}_{i,j}^n)^3$$

$$+ \chi \big(\big(1 + (\hat{\varphi}_{i,j}^n)^2 \big) - 2\chi \hat{\varphi}_{i,j}^n \hat{\varphi}_{i,j}^{n+1/2} \big)(\theta_{i,j}^k - \bar{\theta}_0), \tag{4.60}$$

where $i = 1, \ldots, I - 1$, $j = 0, \ldots, J$, and $n > 0$. The second equation in this system becomes

$$\frac{\hat{\varphi}_{i,j}^{n+1} - \hat{\varphi}_{i,j}^{n+1/2}}{\tau} - \varepsilon \hat{\beta}^2 \frac{1}{r_i^2} \frac{1}{\sin \vartheta_j} \frac{1}{2h_\vartheta^2} \Big[\big(\sin \vartheta_{j+1} + \sin \vartheta_j \big)(\hat{\varphi}_{i,j+1}^{n+1} - \hat{\varphi}_{i,j}^{n+1})$$

$$- \big(\sin \vartheta_j + \sin \vartheta_{j-1} \big)(\hat{\varphi}_{i,j}^{n+1} - \varphi_{i,j-1}^{n+1}) \Big] = 0, \tag{4.61}$$

where $i = 0, \ldots, I$, $j = 1, \ldots, J - 1$, and $n > 0$.

Equation (4.60) is supplemented with the difference boundary conditions

$$\varphi_{0,j}^{n+1/2} = \varphi_{1,j}^{n+1/2}; \quad \varphi_{I,j}^{n+1/2} = \varphi_{I-1,j}^{n+1/2}, \tag{4.62}$$

and Eq. (4.61) is supplemented with the boundary conditions

$$\varphi_{i,0}^{n+1} = \varphi_{i,1}^{n+1}; \quad \varphi_{i,J}^{n+1} = \varphi_{i,J-1}^{n+1}. \tag{4.63}$$

Table 4.2 Free boundary position constructed by condition (4.11) (r_{sgt}) and constructed by using the numerical solution (r_{num}) at a given temperature

$\theta - \bar{\theta}_0$	r_{sgt}	r_{num}
-0.3	0.0089	0.0091
-0.2	0.0133	0.0132
-0.1	0.0266	0.0265
-0.01	0.266	0.267

These conditions (4.62) and (4.63) are obtained from the Neumann conditions which supplement Eq. (4.37).

We solve the obtained difference problem (4.59)–(4.63) (using the tridiagonal matrix algorithm) until the following stationary condition is satisfied:

$$\left| \hat{\varphi}_{i,j}^{n_2^*} - \hat{\varphi}_{i,j}^{m} \right| < \tilde{\varepsilon}_2 \quad \text{for } \forall\, m > n_2^*,$$

where $\tilde{\varepsilon}_2$ is a prescribed accuracy.

As a result of this operation, we obtain the function

$$\varphi_{i,j}^{k+1} = \hat{\varphi}_{i,j}^{n_2^*} \tag{4.64}$$

at the $(k+1)$th "global" time step.

To verify whether the difference scheme is well conditioned, we must be sure that there exists a stationary solution for the order function at a given temperature. It turns out that such a solution exists and can be constructed by the pseudotime method. A stationary solution appears when the free boundary attains the position at which the Stefan–Gibbs–Thomson condition (4.11) is satisfied. We do not prove this fact analytically but demonstrate its validity by numerical experiments whose results are presented in Table 4.2.

4.4.3 Difference Scheme for the Heat Equation

Now we construct the difference approximation for the heat equation (4.36)

$$\frac{\partial \theta}{\partial t} - \hat{k}\left(A_1 + A_2\right)\theta = -\frac{1}{2}\frac{\partial \varphi}{\partial t} + \hat{F}.$$

It is approximated by the system of difference equations

$$\frac{\theta_{i,j}^{k+1/2}-\theta_{i,j}^{k}}{h_t}-\frac{\hat{k}}{r_i^2}\frac{1}{2h_r^2}\left[\left(r_{i+1}^2+r_i^2\right)^2\left(\theta_{i+1,j}^{k+1/2}-\theta_{i,j}^{k+1/2}\right)-\left(r_i^2+r_{i-1}^2\right)^2\left(\theta_{i,j}^{k+1/2}-\theta_{i-1,j}^{k+1/2}\right)\right]$$

$$=\frac{1}{2}\frac{\varphi_{i,j}^{k+1}-\varphi_{i,j}^{k}}{h_t}+\hat{F}_{i,j}^{k},\tag{4.65}$$

where $i = 1, \ldots, I-1$, $j = 0, \ldots, J$, and $n > 0$,

$$\frac{\theta_{i,j}^{k+1}-\theta_{i,j}^{k+1/2}}{h_t}-\frac{\hat{k}}{r_i^2}\frac{1}{2h_\vartheta^2}\frac{1}{\sin\vartheta_j}\left[\left(\sin\vartheta_{j+1}+\sin\vartheta_j\right)\left(\theta_{i,j+1}^{k+1}-\theta_{i,j}^{k+1}\right)\right.$$

$$\left.-\left(\sin\vartheta_j+\sin\vartheta_{j-1}\right)\left(\theta_{i,j}^{k+1}-\theta_{i,j}^{k+1}\right)\right]=0,\tag{4.66}$$

where $i = 0, \ldots, I$, $j = 1, \ldots, J-1$, and $n > 0$.

In the right-hand side of Eq. (4.65), the term $\hat{F}_{i,j}^{k}$ becomes

$$\hat{F}_{i,j}^{k}=(t_0/l\rho)F_{i,j}^{k},$$

where $F_{i,j}^{k}$ is defined by formula (4.57) and $\varphi_{i,j}^{k+1}$ is determine by solving the difference problem (4.59)–(4.63).

Equation (4.65) is supplemented with the boundary condition at the top of the cathode (4.34) which has the difference form

$$\frac{\theta_{1,j}^{k+1/2}-\theta_{0,j}^{k+1/2}}{h_r}=\frac{c}{l}\frac{r_0}{\lambda}\frac{\left(j_{\mathrm{em}}\right)_j^k}{e}\mathcal{E}_j^k,\tag{4.67}$$

and the boundary condition on the lower base (4.14) which in difference form becomes

$$\theta_{I,j}^{k+1/2}-\theta_{I-1,j}^{k+1/2}=-h_r\alpha_{\mathrm{cool}}.\tag{4.68}$$

Formulas for j_{em} and $\mathcal{E}$ are described in detail in Sect. 4.3.

Equation (4.66) is supplemented with boundary conditions (4.15) which in difference form become

$$\theta_{i,0}^{k+1}=\theta_{i,1}^{k+1};\quad \theta_{i,J}^{k+1}=\theta_{i,J-1}^{k+1}.\tag{4.69}$$

We note that the introduction of additional fictitious time steps at each stage of computations permits numerically solving phase field system (4.36), (4.37) in a very wide range of the model parameters, including the case where the solution enters the stationary regime.

We use the following algorithm for numerical modeling. At the initial time ($k = 0$), the functions θ^0 and φ^0 are given, and the function Ψ^0 is determined by solving the system of corresponding difference equations. Further, at the next time step, i.e., at the $(k + 1)$th step, the algorithm is the following one: in the set of corresponding difference problems, we first find φ^{k+1}, then θ^{k+1}, and finally, Ψ^{k+1}. For details, see Sect. 4.5.

Moreover, we introduce a liquid phase nucleus in this algorithm, which is described in detail in Sect. 4.8 and [3].

4.4.4 Stability of the Difference Scheme

We verify the stability of the above-constructed difference scheme. Generally speaking, it is a very nontrivial problem to verify whether a difference scheme for nonlinear equations is stable. Here we consider only the conditions that guarantee the possibility of using the tridiagonal matrix algorithm for solving the obtained difference equations. We have the following assertion.

Lemma 4.1 *The tridiagonal matrix algorithm is stable (i.e., it can be used to solve the above-obtained difference equations) if the following condition is satisfied:*

$$\tau \leq \frac{\varepsilon}{1 + 2\chi \left| \theta_{i,j}^k - \bar{\theta}_0 \right|}. \tag{4.70}$$

Proof As is known [11], for the formulas of tridiagonal matrix algorithm

$$\mathcal{A}_i y_{i-1} - C_i y_i + \mathcal{B}_i y_{i+1} = -\mathcal{F}_i,$$

$$y_0 = \kappa_1 y_1 + \mu_1; \quad y_N = \kappa_2 y_{N-1} + \mu_2 \tag{4.71}$$

to be stable, it suffices to satisfy the maximum principle conditions:

$$A_i > 0, \ B_i > 0, \ C_i \geqslant A_i + B_i, \tag{4.72}$$

$$0 \leq \kappa_1 \leq 1, \ 0 \leq \kappa_2 \leq 1, \ 0 \leq \kappa_1 + \kappa_2 \leq 2. \tag{4.73}$$

We rewrite the above-obtained difference equations in the form (4.71), and then verify conditions (4.72) and (4.73).

1. The difference equations for the potential (4.49) look as follows:

$$\widetilde{\mathcal{A}}_i^{\hat{\Psi}} \hat{\Psi}_{i-1,j}^{n+1/2} - \widetilde{C}_i^{\hat{\Psi}} \hat{\Psi}_{i,j}^{n+1/2} + \widetilde{\mathcal{B}}_i^{\hat{\Psi}} \hat{\Psi}_{i+1,j}^{n+1/2} = -\widetilde{\mathcal{F}}_i^{\hat{\Psi}}, \tag{4.74}$$

where

$$\widetilde{\mathcal{A}}_i^{\hat{\psi}} = \frac{1}{2h_r^2 r_i^2}\left(\sigma_{i-1,j}^k r_{i-1}^2 + \sigma_{i,j}^k r_i^2\right), \tag{4.75}$$

$$\widetilde{\mathcal{B}}_i^{\hat{\psi}} = \frac{1}{2h_r^2 r_i^2}\left(\sigma_{i,j}^k r_i^2 + \sigma_{i+1,j}^k r_{i+1}^2\right), \tag{4.76}$$

$$\widetilde{C}_i^{\hat{\psi}} = \frac{1}{\tau} + \frac{1}{2h_r^2 r_i^2}\left(\sigma_{i-1,j}^k r_{i-1}^2 + 2\sigma_{i,j}^k r_i^2 + \sigma_{i+1,j}^k r_{i+1}^2\right), \tag{4.77}$$

$$\widetilde{\mathcal{F}}_i^{\hat{\psi}} = \frac{\hat{\Psi}_{i,j}^n}{\tau}. \tag{4.78}$$

Obviously, $\widetilde{\mathcal{A}}_i^{\hat{\psi}} > 0$, $\widetilde{\mathcal{B}}_i^{\hat{\psi}} > 0$ and $\widetilde{C}_i^{\hat{\psi}} = 1/\tau + \widetilde{\mathcal{A}}_i^{\hat{\psi}} + \widetilde{\mathcal{B}}_i^{\hat{\psi}} > \widetilde{\mathcal{A}}_i^{\hat{\psi}} + \widetilde{\mathcal{B}}_i^{\hat{\psi}} > 0$, i.e., condition (4.72) is satisfied.

The boundary conditions for Eqs. (4.49) become (see (4.51), (4.52))

$$\hat{\Psi}_{0,j}^{n+1/2} = 0, \quad \hat{\Psi}_{I,j}^{n+1/2} = \text{const.}$$

Obviously, conditions (4.73) are satisfied (the coefficients κ_1 and κ_2 are equal to zero).

Therefore, the tridiagonal matrix algorithm can be used to solve this difference problem without any restrictions.

2. The difference equations for the potential (4.50) have the form

$$\overline{\mathcal{A}}_j^{\hat{\psi}} \hat{\Psi}_{i,j-1}^{n+1} - \overline{C}_j^{\hat{\psi}} \hat{\Psi}_{i,j}^{n+1} + \overline{\mathcal{B}}_j^{\hat{\psi}} \hat{\Psi}_{i,j+1}^{n+1} = -\overline{\mathcal{F}}_j^{\hat{\psi}}, \tag{4.79}$$

where

$$\overline{\mathcal{A}}_j^{\hat{\psi}} = \frac{1}{2h_\vartheta^2 r_i^2 \sin\vartheta_j}\left(\sigma_{i,j-1}^k \sin\vartheta_{j-1} + \sigma_{i,j}^k \sin\vartheta_j\right), \tag{4.80}$$

$$\overline{\mathcal{B}}_j^{\hat{\psi}} = \frac{1}{2h_r^2 r_i^2 \sin\vartheta_j}\left(\sigma_{i,j}^k \sin\vartheta_j + \sigma_{i,j+1}^k \sin\vartheta_{j+1}\right), \tag{4.81}$$

$$\overline{C}_j^{\hat{\psi}} = \frac{1}{\tau} + \frac{1}{2h_r^2 r_i^2 \sin\vartheta_j}\left(\sigma_{i,j-1}^k \sin\vartheta_{j-1}\right.$$
$$\left. + 2\sigma_{i,j}^k \sin\vartheta_j + \sigma_{i,j+1}^k \sin\vartheta_{j+1}\right), \tag{4.82}$$

$$\overline{\mathcal{F}}_j^{\hat{\psi}} = \frac{\hat{\Psi}_{i,j}^{n+1/2}}{\tau}. \tag{4.83}$$

Obviously, $\overline{\mathcal{A}}_j^{\hat{\psi}} > 0$, $\overline{\mathcal{B}}_j^{\hat{\psi}} > 0$ and $\overline{C}_j^{\hat{\psi}} = 1/\tau + \overline{\mathcal{A}}_j^{\hat{\psi}} + \overline{\mathcal{B}}_j^{\hat{\psi}} > \overline{\mathcal{A}}_j^{\hat{\psi}} + \overline{\mathcal{B}}_j^{\hat{\psi}} > 0$, i.e., condition (4.72) is satisfied.

The boundary conditions for Eqs. (4.50) become (see (4.53))

$$\hat{\Psi}_{i,0}^{n+1} = \hat{\Psi}_{i,1}^{n+1}; \quad \hat{\Psi}_{i,J}^{n+1} = \hat{\Psi}_{i,J-1}^{n+1}.$$

Obviously, conditions (4.73) are satisfied (the coefficients κ_1 and κ_2 are equal to 1).

Therefore, the tridiagonal matrix algorithm can be used to solve this difference problem without any restrictions.

3. The difference equations for the order function (4.60) become

$$\widetilde{\mathcal{A}}_i^{\hat{\varphi}}\,\hat{\varphi}_{i-1,j}^{n+1/2} - \widetilde{C}_i^{\hat{\varphi}}\,\hat{\varphi}_{i,j}^{n+1/2} + \widetilde{\mathcal{B}}_i^{\hat{\varphi}}\,\hat{\varphi}_{i+1,j}^{n+1/2} = -\widetilde{\mathcal{F}}_i^{\hat{\varphi}},$$

where

$$\widetilde{\mathcal{A}}_i^{\hat{\varphi}} = \frac{\varepsilon\hat{\beta}^2}{2h_r^2 r_i^2}\left(r_{i-1}^2 + r_i^2\right),$$

$$\widetilde{\mathcal{B}}_i^{\hat{\varphi}} = \frac{\varepsilon\hat{\beta}^2}{2h_r^2 r_i^2}\left(r_i^2 + r_{i+1}^2\right),$$

$$\widetilde{C}_i^{\hat{\varphi}} = \frac{1}{\tau} + \frac{\varepsilon\hat{\beta}^2}{2h_r^2 r_i^2}\left(r_{i-1}^2 + 2r_i^2 + r_{i+1}^2\right) - \frac{1}{\varepsilon}\left(1 - 3\left(\hat{\varphi}_{i,j}^n\right)^2\right) + 2\chi\hat{\varphi}_{i,j}^n\left(\theta_{i,j}^k - \bar{\theta}_0\right),$$

$$\widetilde{\mathcal{F}}_i^{\hat{\varphi}} = -\frac{\hat{\varphi}_{i,j}^n}{\tau} + \frac{2}{\varepsilon}\left(\hat{\varphi}_{i,j}^n\right)^3 + \chi\left(1 + \left(\hat{\varphi}_{i,j}^n\right)^2\right)\left(\theta_{i,j}^k - \bar{\theta}_0\right).$$

Obviously, $\widetilde{\mathcal{A}}_i^{\hat{\varphi}} > 0$ and $\widetilde{\mathcal{B}}_i^{\hat{\varphi}} > 0$. The coefficient $\widetilde{C}_i^{\hat{\varphi}}$ has the form

$$\widetilde{C}_i^{\hat{\varphi}} = \frac{1}{\tau} + \widetilde{\mathcal{A}}_i^{\hat{\varphi}} + \widetilde{\mathcal{B}}_i^{\hat{\varphi}} - \frac{1}{\varepsilon}\left(1 - 3\left(\hat{\varphi}_{i,j}^n\right)^2\right) + 2\chi\hat{\varphi}_{i,j}^n\left(\theta_{i,j}^k - \bar{\theta}_0\right).$$

Therefore, to satisfy the condition $\widetilde{C}_i^{\hat{\varphi}} \geq \widetilde{\mathcal{A}}_i^{\hat{\varphi}} + \widetilde{\mathcal{B}}_i^{\hat{\varphi}}$, it is necessary to satisfy

$$\frac{1}{\tau} - \frac{1}{\varepsilon}\left(1 - 3\left(\hat{\varphi}_{i,j}^n\right)^2\right) + 2\chi\hat{\varphi}_{i,j}^n\left(\theta_{i,j}^k - \bar{\theta}_0\right) \geq 0,$$

whence we obtain

$$\tau \leq \frac{\varepsilon}{\left(1 - 3\left(\hat{\varphi}_{i,j}^n\right)^2\right) + 2\varepsilon\chi\hat{\varphi}_{i,j}^n\left(\theta_{i,j}^k - \bar{\theta}_0\right)}.$$

Since the function $\hat{\varphi}_{i,j}^n$ ranges in $[-1, 1]$, we obtain the desired condition (4.70)

$$\tau \leq \frac{\varepsilon}{1 + 2\varepsilon\chi\left|\theta_{i,j}^k - \bar{\theta}_0\right|}.$$

The coefficients $\tilde{\kappa}_1 = 1$ and $\tilde{\kappa}_2 = 1$ in boundary conditions (4.62) for Eqs. (4.60) satisfy requirements (4.73).

Therefore, for this difference problem, the tridiagonal matrix algorithm can be used only if we choose the system parameters satisfying condition (4.70).

4. Equations (4.61) can be written as

$$\overline{\mathcal{A}}_j^{\hat{\varphi}}\,\hat{\varphi}_{i,j-1}^{n+1} - \overline{C}_j^{\hat{\varphi}}\,\hat{\varphi}_{i,j}^{n+1} + \overline{\mathcal{B}}_j^{\hat{\varphi}}\,\hat{\varphi}_{i,j+1}^{n+1} = -\overline{\mathcal{F}}_j^{\hat{\varphi}},$$

where

$$\overline{\mathcal{A}}_j^{\hat{\varphi}} = \frac{\varepsilon\hat{\beta}^2}{2h_\vartheta^2 r_i^2 \sin\vartheta_j}\left(\sin\vartheta_{j-1} + \sin\vartheta_j\right),$$

$$\overline{\mathcal{B}}_j^{\hat{\varphi}} = \frac{\varepsilon\hat{\beta}^2}{2h_\vartheta^2 r_i^2 \sin\vartheta_j}\left(\sin\vartheta_j + \sin\vartheta_{j+1}\right),$$

$$\overline{C}_j^{\hat{\varphi}} = \frac{1}{\tau} + \frac{\varepsilon\hat{\beta}^2}{2h_\vartheta^2 r_i^2 \sin\vartheta_j}\left(\sin\vartheta_{j-1} + 2\sin\vartheta_j \sin\vartheta_{j+1}\right),$$

$$\overline{\mathcal{F}}_j^{\hat{\varphi}} = \frac{\hat{\varphi}_{i,j}^{n+1/2}}{\tau}.$$

Similarly, it is obvious that $\overline{\mathcal{A}}_j > 0$ and $\overline{\mathcal{B}}_j > 0$. It is also obvious that $\overline{C}_j = 1/\tau + \overline{\mathcal{A}}_j + \overline{\mathcal{B}}_j > \overline{\mathcal{A}}_j + \overline{\mathcal{B}}_j$, i.e., condition (4.72) is always satisfied.

The coefficients $\bar{\kappa}_1 = 1$ and $\bar{\kappa}_2 = 1$ in boundary conditions (4.63) for Eq. (4.61) satisfy requirements (4.73).

Therefore, for this difference problem, the tridiagonal matrix algorithm can be used without restrictions.

5. Now we consider the equation for the temperature. Difference equations (4.65) become

$$\widetilde{\mathcal{A}}_i^\theta\,\theta_{i-1,j}^{k+1/2} - \widetilde{C}_i^\theta\,\theta_{i,j}^{k+1/2} + \widetilde{\mathcal{B}}_i^\theta\,\theta_{i+1,j}^{k+1/2} = -\widetilde{\mathcal{F}}_i^\theta,$$

where

$$\widetilde{\mathcal{A}}_i^\theta = \frac{\hat{k}}{2h_r^2 r_i^2}\left(r_{i-1}^2 + r_i^2\right),$$

$$\widetilde{\mathcal{B}}_i^\theta = \frac{\hat{k}}{2h_r^2 r_i^2}\left(r_i^2 + r_{i+1}^2\right),$$

$$\widetilde{C}_i^\theta = \frac{1}{h_t} + \frac{\hat{k}}{2h_r^2 r_i^2}\left(r_{i-1}^2 + 2r_i^2 + r_{i+1}^2\right),$$

$$\widetilde{\mathcal{F}}_i^\theta = -\frac{\hat{\theta}_{i,j}^k}{h_t} - \frac{1}{2}\frac{\varphi_{i,j}^{k+1} - \varphi_{i,j}^k}{h_t} + \hat{F}_{i,j}^k.$$

Obviously, $\widetilde{\mathcal{A}}_i^{\hat{\varphi}} > 0$, $\widetilde{\mathcal{B}}_i^{\hat{\varphi}} > 0$ and $\widetilde{C}_i^\theta = 1/h_t + \widetilde{\mathcal{A}}_i^\theta + \widetilde{\mathcal{B}}_i^\theta > \widetilde{\mathcal{A}}_i^\theta + \widetilde{\mathcal{B}}_i^\theta$.

The coefficients $\tilde{\kappa}_1' = 1$ in boundary condition (4.67) and $\tilde{\kappa}_2' = 1$ in boundary condition (4.68) satisfy conditions (4.73).

Therefore, for this difference problem, the tridiagonal matrix algorithm can be used without restrictions.

6. Equations (4.66) become

$$\overline{\mathcal{A}}_j^{\theta} \theta_{i,j-1}^{k+1} - \overline{C}_j^{\theta} \theta_{i,j}^{k+1} + \overline{\mathcal{B}}_j^{\theta} \theta_{i,j+1}^{k+1} = -\overline{\mathcal{F}}_j^{\theta},$$

where

$$\overline{\mathcal{A}}_j^{\theta} = \frac{\hat{k}}{2h_{\vartheta}^2 r_i^2 \sin \vartheta_j} \left(\sin \vartheta_{j-1} + \sin \vartheta_j \right),$$

$$\overline{\mathcal{B}}_j^{\theta} = \frac{\hat{k}}{2h_{\vartheta}^2 r_i^2 \sin \vartheta_j} \left(\sin \vartheta_j + \sin \vartheta_{j+1} \right),$$

$$\overline{C}_j^{\theta} = \frac{1}{h_t} + \frac{\hat{k}}{2h_{\vartheta}^2 r_i^2 \sin \vartheta_j} \left(\sin \vartheta_{j-1} + 2 \sin \vartheta_j \sin \vartheta_{j+1} \right),$$

$$\overline{\mathcal{F}}_j^{\theta} = \frac{\theta_{i,j}^{k+1/2}}{h_t}.$$

Similarly, it is clear that conditions (4.72) for the coefficients $\overline{\mathcal{A}}_j^{\theta}$, $\overline{\mathcal{B}}_j^{\theta}$, and $\overline{C}_j^{\theta}$ are satisfied. The coefficients $\bar{\kappa}_1' = 1$ and $\bar{\kappa}_2' = 1$ in boundary conditions (4.69) for Eq. (4.66) satisfy conditions (4.73).

Therefore, for this difference problem, the tridiagonal matrix algorithm can be used without restrictions.

Thus, the formulas of the tridiagonal matrix algorithm are stable for any parameters of the original problem satisfying condition (4.70). $\qquad\square$

Remark 4.6 In the calculations, the parameter is $\varepsilon \sim 10^{-2}$, the coefficient χ is equal to $\sqrt{2}/2$, and the quantity $|\theta_{i,j}^k - \bar{\theta}_0|$ does not exceed 10. Thus, condition (4.70) can be simplified to the condition

$$\tau \leqslant \frac{\varepsilon}{2}. \tag{4.84}$$

4.4.5 One More Version of the Difference Scheme

In conclusion, we consider another difference scheme for the phase field system which can be used to solve the problem numerically. This scheme will be called the difference scheme 2. The new scheme differs from the old one only in a new difference equation for the radial part of heat equation (4.36) and the corresponding boundary conditions. In turn, the change $w = \theta/r$ takes the radial part of the heat equation

$$\frac{\partial \theta}{\partial t} - \hat{k}\frac{1}{r^2}\frac{\partial}{\partial r}\left(r^2\frac{\partial \theta}{\partial r}\right) = -\frac{1}{2}\frac{\partial \varphi}{\partial t} + \hat{F}$$

to the heat equation with constant coefficients

$$\frac{\partial w}{\partial t} - \hat{k}\frac{\partial^2 w}{\partial r^2} = -\frac{1}{2}r\frac{\partial \varphi}{\partial t} + r\hat{F}.$$

We write this equation in the form of finite differences:

$$\frac{w_{i,j}^{k+1/2} - w_{i,j}^{k}}{\tau} = \hat{k}\frac{w_{i-1,j}^{k+1/2} - 2w_{i,j}^{k+1/2} + w_{i+1,j}^{k+1/2}}{h^2} - \frac{1}{2}\frac{\varphi_{i,j}^{k+1} - \varphi_{i,j}^{k}}{\tau} + r_i\hat{F}_{i,j}^{k}, \quad (4.85)$$

where $i = 1, \ldots, I - 1$, $j = 0, \ldots, J$, and $k > 0$.

The new difference scheme is obtained from the scheme considered above, where Eq. (4.65) is replaced by Eq. (4.85) and boundary conditions (4.67) and (4.68) are respectively replaced by (4.86) and (4.87), so that they have the form

$$R_0\frac{w_{1,j}^{k+1/2} - w_{0,j}^{k+1/2}}{h_r} = \frac{c}{l}\frac{r_0}{\lambda}\frac{j_{\text{em}}^{k}}{e}\mathcal{E}_j^{k}, \quad (4.86)$$

$$R_0\frac{w_{I,j}^{k+1/2} - w_{I-1,j}^{k+1/2}}{h_r} = -\alpha_{\text{cool}}. \quad (4.87)$$

As previously, the new difference scheme can be solved numerically by the tridiagonal matrix algorithm; when the tridiagonal matrix algorithm is applied in the variable r for the heat equation, the values of θ are replaced by w, and then the calculated values of w are conversely replaced by θ.

To verify whether the difference scheme 2 is well conditioned, we solve the problem

$$\begin{cases} \dfrac{\partial \theta}{\partial t} = \hat{k}\Delta\theta + \exp\left(\gamma(r_0 - r)\right), \\ \dfrac{\partial \theta}{\partial r}\bigg|_{r=R_0} = \alpha\theta, \\ \dfrac{\partial \theta}{\partial r}\bigg|_{r=R} = 0. \end{cases} \quad (4.88)$$

Here the right-hand side and the boundary condition are similar to the corresponding data in (4.36) and (4.37). The exact stationary solution can easily be found for problem (4.88):

$$\theta(r) = \frac{1}{kr R_0^2 \alpha \gamma^3}e^{-(r+R)\gamma}\big(-e^{(R+R_0)\gamma}R_0^2\alpha(2 + r\gamma)$$
$$- e^{(r+R_0)\gamma}(r + rR_0\alpha - R_0^2\alpha)(2 + 2R\gamma + R^2\gamma^2)$$
$$+ e^{(r+R)\gamma}r(2 + 2R_0(\alpha + \gamma) + R_0^2\gamma(\alpha + \gamma))\big).$$

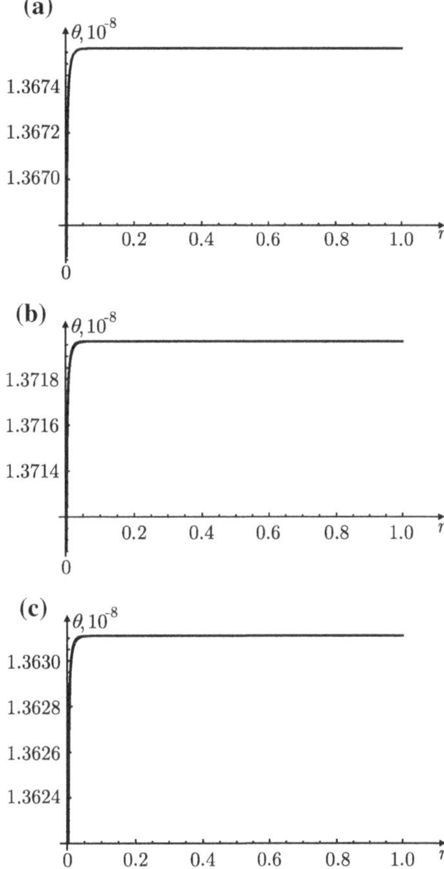

Fig. 4.2 Exact solution and numerical solutions of model problem (4.88): **a** exact solution; **b** numerical solution obtained by the main difference scheme with the time step $\tau = 0.0003$; **c** numerical solution obtained by difference scheme 2 with the time step $\tau = 0.0005$

Now we compare the exact solution with numerical solutions obtained for the following parameters:

$$\hat{k} = 10^8, \ \gamma = 100, \ R_0 = 0.0025, \ \alpha = 0.3, \ R = 1, \ h_r = 10^{-5}.$$

In what follows, Fig. 4.2 shows the graphs of stationary solutions: the exact solution, the solution calculated by using the difference scheme described in this section (the main scheme which was used to obtain all results presented below), and the solution calculated by difference scheme 2. One can see that both difference schemes give good approximations to the exact solution.

4.4.6   Choice of Steps of the Difference Scheme

Now we discuss the choice of optimal steps h_r and τ. Despite the fact that the equations are here approximated by implicit difference schemes, the system of equations is unstable, and hence the proof of stability and convergence is a very nontrivial problem. The calculations based on implicit difference schemes do not give a correct result for all steps. To choose appropriate parameters of the difference scheme (steps), we use the balance relations obtained by integrating the system of Eqs. (4.12), (4.13) over the cathode volume V. As will be noted below in Sect. 4.7, the temperature and the order function are independent of the angles ϕ and ϑ under the initial and boundary conditions used in our calculations, and hence the integral over the volume V becomes an integral over the coordinate:

$$\iiint\limits_V f(r,t)\mathrm{d}V = \int_{R_0}^R r^2 f(r,t)\mathrm{d}r \cdot \int_0^\pi \mathrm{d}\phi \cdot \int_{-\Theta}^{\Theta} |\sin\vartheta|\mathrm{d}\vartheta$$

$$= 2\pi(1-\cos\Theta)\int_{R_0}^R r^2 f(r,t)\mathrm{d}r.$$

The balance relation for heat equation (4.12) becomes

$$\frac{\mathrm{d}}{\mathrm{d}t}\int_{R_0}^R r^2(\theta-\bar\theta_0)\,\mathrm{d}r = \hat k R^2 \left.\frac{\partial\theta}{\partial r}\right|_{r=R} - \hat k R_0^2 \left.\frac{\partial\theta}{\partial r}\right|_{r=R_0}$$

$$-\frac{1}{2}\int_{R_0}^R r^2\frac{\partial\varphi}{\partial t}\mathrm{d}r + \int_{R_0}^R r^2\hat F\mathrm{d}r. \tag{4.89}$$

For phase field equation (4.13), the balance relation becomes

$$\int_{R_0}^R \varepsilon\hat\alpha r^2\frac{\partial\varphi}{\partial t}\mathrm{d}r - \int_{R_0}^R \varepsilon\hat\beta\frac{\partial}{\partial r}r^2\frac{\partial\varphi}{\partial r}\mathrm{d}r$$

$$= \int_{R_0}^R \frac{1}{\varepsilon}r^2\big(\varphi-\varphi^3\big)\mathrm{d}r + \int_{R_0}^R \chi r^2\big(\theta-\bar\theta_0\big)\mathrm{d}r. \tag{4.90}$$

Balance relations (4.89), (4.90) were calculated by numerical integration at a temperature close to the melting temperature $\bar\theta_0$. It turned out that these relations strongly depend on the steps of the difference scheme, and they are most precisely satisfied for the steps

$$h_r = 10^{-4}, \quad \tau = 10^{-5}. \tag{4.91}$$

Therefore, these steps are used in the further calculations described in the book.

4.5   Algorithm for Solving the Difference Equations and Possible Versions of its Parallelization

The difference scheme given in Sect. 4.4 can be solved using computer by the usual tridiagonal matrix algorithm [11]. Namely, the following algorithm is used. For each $(k + 1)$st time step, the following calculations are performed.

Algorithm A1 (calculation of one (k + 1st) time iteration for the complete system of equations)

A1.1. We determine the order function φ^{k+1} using the temperature value θ^k calculated at the kth step. For this, it is necessary to solve difference equations (4.60), (4.61) with appropriate boundary conditions, see Sect. 4.4.2.
A1.2. We determine the potential Ψ^{k+1} similarly using the temperature value θ^k obtained at the preceding step. For this, it is necessary to solve difference equations (4.49), (4.50) with corresponding boundary conditions, see Sect. 4.4.1. Then we determine the value of the heat release capacity density F^{k+1} by formula (4.57).
A1.3 We determines the temperature θ^{k+1} using the calculated values of the functions φ^{k+1} and F^{k+1}. For this, it is necessary to solve difference equations (4.65), (4.66) with corresponding boundary conditions, see Sect. 4.4.3.

We note that, at all steps of algorithm A1, the difference equations are solved by the same method, namely, as was already noted, by the tridiagonal matrix algorithm. We consider one of the steps of this algorithm in more detail. For example, step A1.2.

For convenience, we introduce the following notation. We denote the set of all grid indices in the variable r by N_r (see (4.45)), and in the variable ϑ, by N_ϑ (see (4.46)):

$$N_r - \{0, 1, 2, \ldots, I\}, \quad N_\vartheta - \{0, 1, 2, \ldots, J\}.$$

Then the following algorithm is used. For all $i \in N_r$, we solve difference equation (4.49) with corresponding boundary conditions. In matrix form, this equation takes the form (4.74), i.e., is a system of linear algebraic equations (SLAE) whose matrix is three-diagonal. Therefore, as was already noted, to obtain the solutions of this SLAE, one can apply the tridiagonal matrix algorithm. Then, for all $j \in N_\vartheta$, it is necessary to solve difference equation (4.50) which, in matrix form (see (4.79)), is similar to SLAE with three-diagonal matrix. Similarly, this system can be solved by the tridiagonal matrix algorithm. We write this algorithm in a more rigorous form.

Algorithm A2 (calculations of step A1.2 of algorithm A1)

A2.1a. For all $j \in N_r$, to use formulas (4.75)–(4.78) to calculate the coefficients $\widetilde{A}_i^{\hat{\Psi}}$, $\widetilde{B}_i^{\hat{\Psi}}$, $\widetilde{C}_i^{\hat{\Psi}}$, and $\widetilde{F}_i^{\hat{\Psi}}$.
A2.1b. To solve the SLAE obtained at the preceding step; the number of equations is $J + 1$, and the dimension of each of them is $I + 1$.

A2.2a. For all $i \in N_\vartheta$, to use formulas (4.80)–(4.83) to calculate the coefficients $\overline{A}_j^{\hat{\psi}}$, $\overline{B}_j^{\hat{\psi}}$, $\overline{C}_j^{\hat{\psi}}$, and $\overline{F}_j^{\hat{\psi}}$.

A2.2b. To solve the SLAE obtained at the preceding step; the number of equations is $I + 1$, and the dimension of each of them is $J + 1$.

After determining the potential, it is necessary trivially to calculate F^{k+1} by the formulas given in Sect. 4.4.1; here we do not consider these calculations.

We note that despite the small size of the cathode area, the obtained solution varies fast, and thus it is necessary to use very small steps (see (4.91)), which means that the number of points of the difference grid is large. In the above notation, this means that the number of elements of the sets N_r and N_ϑ is large. Therefore, our calculation algorithm is very time consuming. The time required to calculate only one step A1.2 for one time iteration is given in Table 4.3. And to calculate the whole problem, it is necessary to perform all three steps of algorithm A1 with rather many iterations. Obviously, all this requires a tremendous amount of time.

But now the multiprocessor computers are widely used. In this connection, we improved our algorithm so that it can simultaneously run on several processors. Note that here we do not analytically analyze its efficiency but only practically demonstrate the obtained acceleration.

The above-cited algorithm A2 explicitly implies the following two ways for constructing a "parallel" algorithm.

Parallel algorithm AP1

We note that, at steps A2.1b and A2.2b, the SLAE is solved by the tridiagonal matrix algorithm which is a certain sequential algorithm. The following idea immediately arises: to replace it by a certain algorithm which permits solving the SLAE simultaneously by using several processors. Such an algorithm exists and is called the "block" tridiagonal matrix algorithm. Briefly, it consists in the following. The extended matrix of SLAE is divided into several different strips whose number is equal to the number of accessible processors. Then the usual tridiagonal matrix algorithm supplemented with mechanisms of synchronization and data exchange between pro-

Table 4.3 Operation time of algorithm A2 depending on the grid dimensions. One core of AMD Opteron 6234 processor (2.4 GHz) was used for calculations

Grid dimensions	Operation time (in seconds)
1215×1280	70.05
2431×2560	281.39
4863×5120	1129.94
9727×10240	4533.13

Table 4.4 Acceleration obtained by using algorithm AP1 on CPU versus the grid dimensions and the number of processors. Two AMD Opteron 6234 processors (12 cores, 2.4 GHz) were used for calculations

Grid dimensions	Obtained acceleration		
	4 CPU	8 CPU	16 CPU
1215×1280	1.320	1.502	1.375
2431×2560	1.323	1.404	1.379
4863×5120	1.315	1.432	1.380
9727×10240	1.390	1.423	1.322

cessors is simultaneously used in each strip. The acceleration[1] gained in this way is given in Table 4.4.

Parallel algorithm AP2

Another approach is based on the fact that the SLAE obtained at step A2.1a are independent of each other with respect to the index $j \in N_\vartheta$. Therefore, they can be solved simultaneously. Let P be the number of accessible processors. We number them in ascending order starting from 1: $p = 1, 2, \ldots, P$. Now we uniformly divide the set of indices N_ϑ into subsets N_ϑ^p:

$$N_\vartheta = \bigcup_{p=1}^{P} N_\vartheta^p.$$

Then each processor with number p makes steps A2.1a and A2.1b of algorithm A2 for $j \in N_\vartheta^p$. Then, when each processor finishes to operate, we similarly calculate steps A2.2a and A2.2b, because the SLAE obtained at step A2.2a are independent of each other with respect to the index $i \in N_r$ (i.e., in this case, it is necessary to spread the set N_r uniformly on the subsets N_r^p). The obtained acceleration is given in Table 4.5.

Obviously, the obtained results (see Tables 4.4 and 4.5) show that the *algorithm AP2 is optimal*.

We also note that currently, along with the "classical" multiprocessor computers, it becomes more and more popular to use graphic accelerators (GPU) for calculations. A bright example of such an approach is the CUDA technology [2]. We also use this technology to perform calculations by algorithm AP2.

---

[1] Here and below, the notion of acceleration means the ratio of the operation time of some parallel algorithm to the operation time of the sequential algorithm A2 on one CPU (see Table 4.3).

Table 4.5 Acceleration obtained by using algorithm AP2 on CPU versus the grid dimensions and the number of processors. Two AMD Opteron 6234 processors (12 cores, 2.4 GHz) were used for calculations

Grid dimensions	Obtained acceleration				
	2 CPU	4 CPU	8 CPU	16 CPU	24 CPU
1215×1280	36.10	67.70	109.16	95.09	75.57
2431×2560	36.83	66.60	79.46	119.60	118.19
4863×5120	35.71	66.85	75.34	124.11	104.88
9727×10240	33.93	65.52	77.23	114.83	127.26

Table 4.6 Operation time of algorithm AP2 on GPU versus the grid dimensions and the type of used variables. The graphic accelerator NVIDIA TESLA C2070 (448 CUDA cores, 6 Gb memory) was used for calculations

Grid dimensions	Operation time (in seconds)	
	float	double
4863×5120	2.32	3.41
9727×10240	16.61	Memory shortage

We note that the speed of computations on graphic accelerators depends on the type of used variables (floating or double accuracy). The time required for calculations depending on the type of variables and the grid dimensions is given in Table 4.6.

One can see that it is most optimal to use GPU to solve this problem, and the computation time is approximately 273 times less than the time of computation by using the sequential algorithm for the grid 9727×10240.

In conclusion, we note that our algorithm exhibits the "saturation" effect, i.e., for some fixed grid dimensions, there exist a number P_{max} of processors to be used at which the operation time of the algorithm is minimal (i.e., the maximal acceleration is observed), and as the number of processors increases, the operation time does not decrease and, moreover, can even increase, see Table 4.5, and also see the diagram in Fig. 4.3.

This can clearly be seen for the grid 1215×1280: the optimal number of processors to be used is eight (the acceleration by 109.16 times is attained), and if the number of processors increases, then the acceleration begins to decrease (in the case of 24 CPU, the obtained acceleration is already by 75.57 times), see Fig. 4.3

This effect can be explained by the fact that the time required for the data distribution over processors, synchronization operation, and other implicit processes strongly increases with the number P of processors, and for $P > P_{max}$, it begins to exceed the time directly spent for calculations. The operation of data distribution over processors strongly depends the technologies of parallel programming used to implement specific programs. We used the MPI (Message Passing Interface) standard, i.e., the most widely used standard of data exchange interface in parallel program-

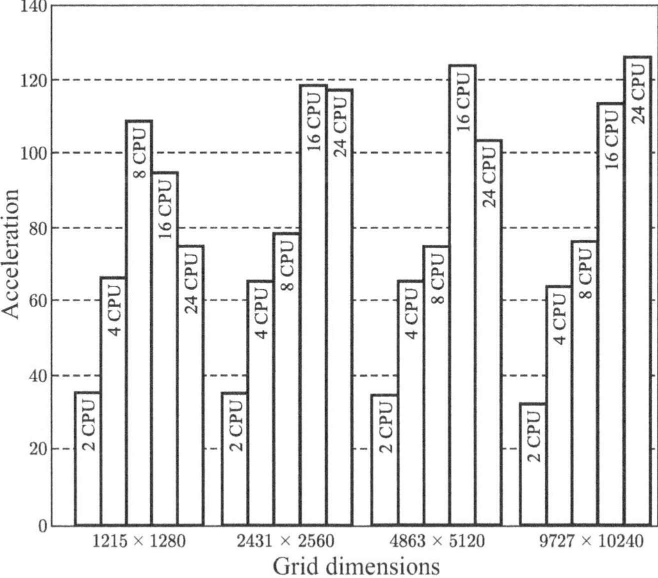

Fig. 4.3 Diagram: the acceleration obtained by algorithm AP2 on CPU versus the grid dimensions and the number of processors. Two AMD Opteron 6234 processors (12 cores, 2.4 GHz) were used for calculations

ming, which has been implemented for rather many computer platforms [10]. Along with this standard, there also exist other technologies, for example, OpenMP (Open Multi—Processing), i.e., an open standard for parallelizing the programs intended for computer platforms with common memory [1, 10].

4.6   Some Remarks About the Calculation of the Electric Potential

In possible real applications (field emission display, TV, and so on), one should consider the following model anode–cathode configuration, see Fig. 4.4. Figure 4.4 demonstrates one "cell" of nearly real configuration with given values of the potential at the cathode bottom ($\Psi = \Psi_2$) and at the anode ($\Psi = \Psi_1$), and l_g is the distance between the top of the cathode and the anode. A conic cathode is placed in a vacuum. By σ_v and σ_c we denote the conductivity of the vacuum and the cathode material (Si in the case under study), $\sigma_v \ll \sigma_c$.

It is well known that the equation for the potential Ψ is a Poisson-type equation

$$\nabla(\sigma(\theta)\nabla\Psi) = R, \tag{4.92}$$

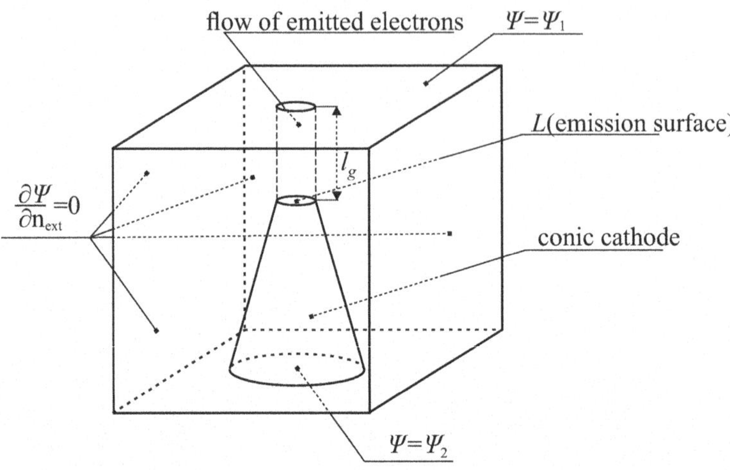

Fig. 4.4 Cathode–anode system

where R is a dimensionless function describing the charge distribution in the "frozen" flow of emitted electrons, and

$$
\sigma =
\begin{cases}
\sigma_v, & \text{in the vacuum domain,} \\
\sigma_c, & \text{in the cathode domain.}
\end{cases}
$$

Let $\mathbf{n}_{\text{ext}}$ and $\mathbf{n}_{\text{int}}$ be the unit outer and inner normals, respectively. We put

$$
\frac{\partial\Psi}{\partial\mathbf{n}_{\text{ext}}} = 0
$$

on the lower boundary of the cell. Let V be the domain of the cell, and let V_{int} be the domain of the cathode, $V_{\text{ext}} = V \setminus V_{\text{int}}$.

Further, we use the definition of a generalized solution of (4.92): for each test function $\zeta(x) \in C^{\infty}(V)$, $\zeta(x) = 0$ at the upper and lower bottoms of the cell, the following integral identity holds:

$$
\iint_V \langle \sigma\nabla\Psi, \nabla\zeta \rangle \, dV + \iint_V R\zeta \, dV = 0. \tag{4.93}
$$

Now we apply definition (4.93) to Eq. (4.92) with discontinuous coefficient σ. Integrating in (4.93) "by parts", we obtain

Table 4.7 Geometric parameters corresponding to Figs. 4.5, 4.6, 4.7 and 4.8

Half-angle at the top of the cathode	12°
Height^a of the "cell"	1
Potential Ψ_2	4 V
Potential Ψ_1	0 V
The gap between the cathode and the anode	0.004

^{a} It is equivalent to 10^{-5} m, see Sect. 4.1

$$
-\sigma_v \int_{V_{\text{ext}}} \zeta \Delta \Psi \, dV - \sigma_c \int_{V \backslash V_{\text{ext}}} \zeta \Delta \Psi \, dV + \sigma_c \int_{\partial (V \backslash V_{\text{ext}})} \zeta \frac{\partial \Psi}{\partial \mathbf{n}_{\text{ext}}} \, dS
$$
$$
+ \sigma_v \int_{\partial V_{\text{ext}}} \zeta \frac{\partial \Psi}{\partial \mathbf{n}_{\text{ext}}} \, dS = - \int_{V_{\text{ext}}} \zeta R \, dV - \int_{V_{\text{int}}} \zeta R \, dV
$$

where $\partial(V \setminus V_{\text{ext}}) = \partial V_{\text{int}}$ is the conic part of the cathode domain boundary (except for the bottom) and dS is the elementary surface area.

This implies the following problem for (4.92):

$$
\sigma_v \Delta \Psi = R \text{ in } V_{\text{ext}},
$$
$$
\sigma_c \Delta \Psi = R \text{ in } V_{\text{int}}, \tag{4.94}
$$

$$
\left[\sigma \frac{\partial \Psi}{\partial \mathbf{n}} \right] = 0 \text{ on } \partial V_{\text{int}},
$$
$$
\frac{\partial \Psi}{\partial \mathbf{n}_{\text{ext}}} = 0 \text{ on } \partial V. \tag{4.95}
$$

(The last relation holds except for the upper and lower bottoms). It is easy to see that, in turn, this implies $\Psi = \Psi_2$ in the cathode domain (with extremely small deviations) and a function which agrees with this constant solution $\Psi_{in} = \Psi_2$ with the boundary conditions on ∂V, see Figs. 4.5, 4.6, 4.7 and 4.8.

The geometric parameters corresponding to Figs. 4.5, 4.6, 4.7 and 4.8 are shown in Table 4.7. The potential in Figs. 4.5, 4.6, 4.7 and 4.8 is measured in Volts.

Finally, this means that the electric tension is too small to generate significant values of an electric current in the cathode domain and the field emission is extremely low in this case. To avoid this contradiction, one should take into account that there is a potential barrier near the emission surface, see Sect. 2.4. Thus, we can assume that there is no normal jump of electric tension in the conic part of the cathode domain, but this is not true for the surface of emission.

Let j_{em} be the density of electrons which form the emission current, and let j_{in} be the current coming to L from inside. At the top of the cathode, we have

$$j_{\text{em}} = D_g j_{\text{in}}, \tag{4.96}$$

where D_g is the transmission coefficient of the potential barrier on the emission surface. If $D_g < 1$, then the condition of absence of the potential normal derivative is not fulfilled. This means that we should investigate the sum of integrals on the emission surface L in more detail:

$$\sigma_c \int_L \zeta \frac{\partial \Psi}{\partial \mathbf{n}_{\text{ext}}} \, \mathrm{d}L + \sigma_v \int_L \zeta \frac{\partial \Psi}{\partial \mathbf{n}_{\text{int}}} \, \mathrm{d}L. \tag{4.97}$$

From (4.96) we have $\frac{\partial \Psi}{\partial \mathbf{n}_{\text{ext}}} = -D_g \frac{\partial \Psi}{\partial \mathbf{n}_{\text{int}}}$, and the sum (4.97) can be expressed as

$$- \sigma_c D_g \int_L \zeta \frac{\partial \Psi}{\partial \mathbf{n}_{\text{int}}} \, \mathrm{d}L + \sigma_v \int_L \zeta \frac{\partial \Psi}{\partial \mathbf{n}_{\text{int}}} \, \mathrm{d}_g L$$

$$= \int_L (\sigma_v - \sigma_c D) \zeta \frac{\partial \Psi}{\partial \mathbf{n}_{\text{int}}} \, \mathrm{d}L. \tag{4.98}$$

The right-hand side of (4.98) can be presented as the action of the generalized function $(\sigma_v - \sigma_c D_g) \frac{\partial \Psi}{\partial \mathbf{n}_{\text{int}}} \delta\big|_L$ on a test function $\zeta(x)$, where $\delta\big|_L$ is the Dirac delta function on the surface L.

So we can put $\sigma_{\text{ext}} = 0$. Thus, we have

$$(\sigma_v - \sigma_c D_g) \frac{\partial \Psi}{\partial \mathbf{n}_{\text{int}}} \delta\big|_L \overset{\text{def}}{=} R_1, \tag{4.99}$$

and for (4.93), we should add $-R_1$ to R.

Now there is a problem of how to calculate R_1. To this end, we recall that the boundary condition on the surface of emission for the heat equation is given by the Nottingham effect (see (4.34)) and has the form

$$\frac{\partial \theta}{\partial \mathbf{n}_{\text{ext}}} = k j_{\text{em}}, \tag{4.100}$$

where k is defined in (4.34) and j_{em} is the emission current density directed towards $\mathbf{n}_{\text{ext}}$, $j_{\text{em}} = D_g \sigma_c E_{\text{int}}$. On the emission surface, we have $E_{\text{int}} = \frac{\partial \Psi}{\partial \mathbf{n}_{\text{ext}}}$. Indeed,

$$\frac{\partial \Psi}{\partial \mathbf{n}_{\text{ext}}} = -\frac{\partial \Psi}{\partial \mathbf{n}_{\text{int}}} = \frac{\partial \theta}{\partial \mathbf{n}_{\text{ext}}} (k \sigma_c D_g)^{-1}. \tag{4.101}$$

The coefficient D_g is in fact the ratio $j_{\text{em}}/j_{\text{in}}$.

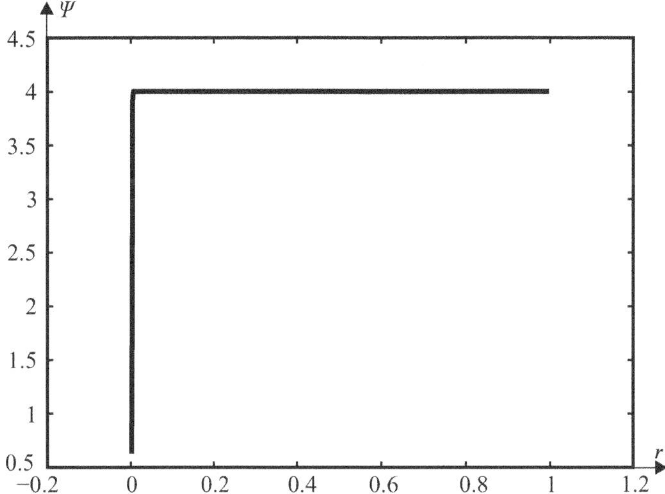

Fig. 4.5 Graph of the potential along the cathode axis (full)

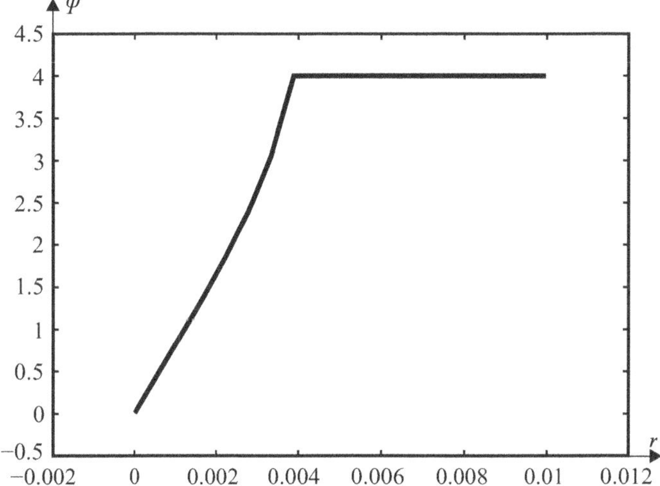

Fig. 4.6 Graph of the potential along the cathode axis (near the top of the cathode)

Our consideration described above can be used to modify the suggested algorithm for calculating the solution because, for given initial values of temperature and potential, one can calculate all quantities required to calculate the potential at the next time step.

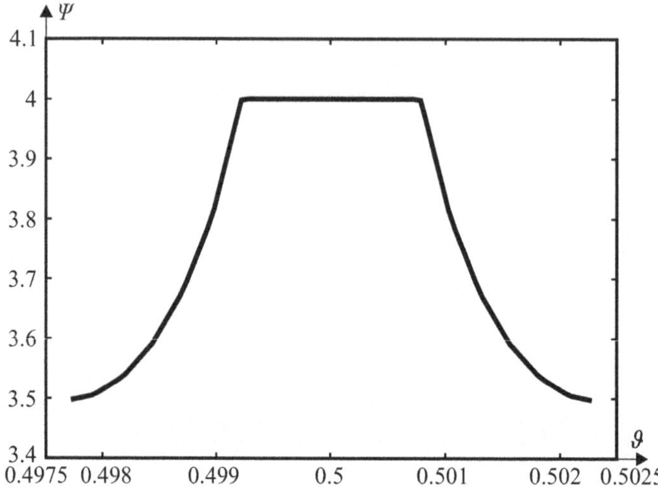

Fig. 4.7 Graph of the potential in the orthogonal direction to the cathode axis, at $r = 0.005$

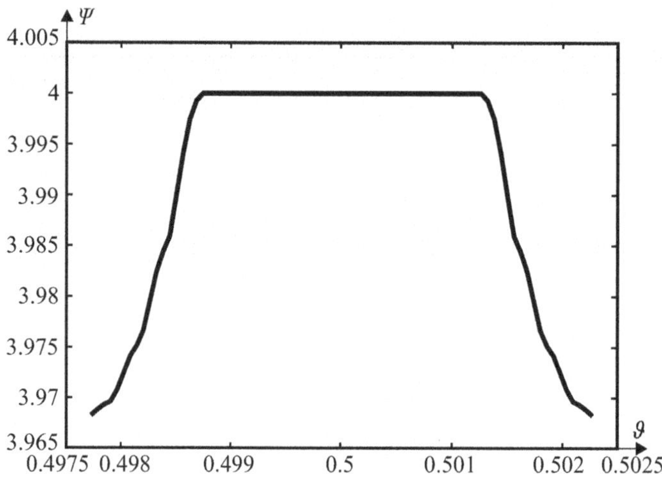

Fig. 4.8 Graph of the potential in the orthogonal direction to the cathode axis, at $r = 0.008$

4.7 Results of Numerical Experiments

In this section, we present graphs of most interesting results of numerical modeling: formation of the liquid phase and solidification. Further, on all graphs in this section, we use the dimensionless temperature θ whose reference point coincides with the dimensionless temperature of the cathode melting $\bar{\theta}_0$, and the measurement unit of dimensionless temperature is equal to 241.89 K.

Table 4.8 The model parameters used to demonstrate the motion of free boundaries

$\hat{k}$	1	$\hat{\alpha}$	1	$\hat{\beta}$	0.03647
ε	0.03	$\bar{\theta}_0$	7.028	R_0	0.01
R	0.25	h_r	0.0001	τ	0.0005

4.7.1 Nonmonotone Behavior of Free Boundaries

In this section, we show a numerical example of nonmonotone behavior of the free boundary in the problem of thermal emission. We note that the parameters used in this section do not correspond to the physical parameters of the cathode considered in this book.

To demonstrate that the behavior of the free boundary is not monotone, we significantly decrease the thermal conductivity (which is similar to the increase in the geometric dimensions), increase the area of the top of the cathode (decrease the action of the Nottingham effect), and increase the influence of the curvature on the deviation of the temperature near the free boundary from the melting temperature. More precisely, the values of the parameters used in this section are given in Table 4.8.

Figure 4.9a illustrates the process of formation of melting regions by graphs of the order function φ before formation of the melting region (the solid phase over the whole region at time t_0, $\varphi = -1$), at time t_1 of origination of the liquid phase nucleus (inside the region, there is an interval on which the order function is greater than zero but less than 1), and after complete formation of the melting region at time t_2 (the order function is equal to 1 in this region). The corresponding graphs of the temperature are shown in Fig. 4.9b.

Figure 4.10 illustrates the whole dynamics of motion of free boundaries up to their coalescence and the disappearance of the liquid phase. The corresponding graphs of the order function are shown in Fig. 4.11. The times t_0, t_1, t_2, and t_3 correspond to the initial data, the liquid phase formation, the time at which the melting region is maximal, and the disappearance of the liquid phase.

The model parameters used to obtain Figs. 4.9, 4.10 and 4.11 are given in Table 4.8, the boundary conditions on the lower base has the form

$$\left.\frac{\partial \theta}{\partial r}\right|_{r=R} = -\theta + 1.24, \tag{4.102}$$

the coefficient in the boundary condition at the top of the cathode is

$$\frac{c}{l}\frac{r_0}{\lambda}\frac{j_{em}}{e} = 300, \tag{4.103}$$

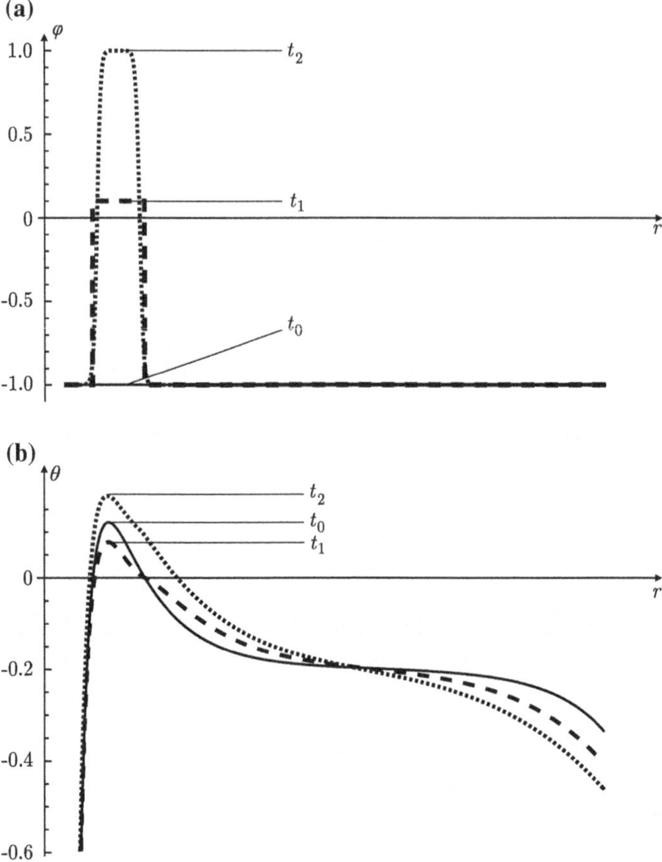

Fig. 4.9 Formation of the melting region. The graphs correspond to the times $t = t_0, t_1, t_2$. At time $t = t_0$, only the solid phase exists; at $t = t_1$, the liquid phase nucleus appears; at $t = t_2$, the melting region is formed

and the right-hand side of the heat equation is here calculated by the simplified expression $F = 0.0022/r^4$ (as compared to the right-hand side of the expression used in calculations with the parameters corresponding to those obtained experimentally in [5]).

4.7.2 Results of Modeling with Physical Parameters Corresponding to Experimental Data

The difficulties arising in the experimental observation of the phenomenon described in the preceding section can also be related to the fact that because of high thermal conductivity of silicon and small dimensions of the cathode, the heat propagates and

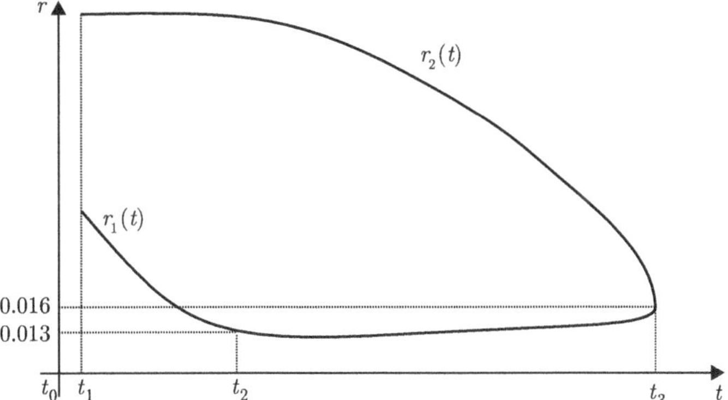

Fig. 4.10 Trajectory of motion of free boundaries. The digits denote a variation in the direction of motion of free boundaries ($t = t_2$) and the coalescence of free boundaries ($t = t_3$), the solidification occurs at this time; r_1 and r_2 are the liquid phase boundaries, the nearest to and the farthest from the top of the cathode, respectively

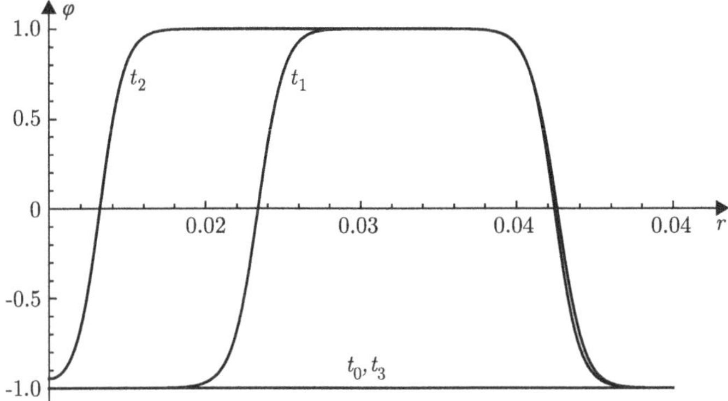

Fig. 4.11 Dynamics of the order function. The digits denote: $t = t_1$ is the time of liquid phase formation; $t = t_2$ is the time at which the free boundaries stop and change the motion direction; $t = t_3$ is the time of the cathode solidification and the coalescence of free boundaries

levels very fast all over the cathode. Therefore, it is very difficult experimentally to fix a significant temperature difference inside the cathode, i.e., to see the boundary of the melting region formation. Namely, the cathode seems to be melting and then becomes completely solid. This is well demonstrated by the temperature graphs obtained by calculating for real values of the parameters at different times $t_0 < t_1 < t_2$ and shown in Fig. 4.12. In this figure, despite the choice of a sufficiently small step, no specific behavior of the temperature near the "top" (upper base) of the cathode can be seen at all. The temperature behaves there like a boundary layer because of the Nottingham effect. The amplitude of this boundary layer is small, a typical graph at $t = t_2$ is shown in Fig. 4.13. The graphs at other times are similar.

Starting from Fig. 4.12 and further, we use the same set of parameters corresponding to the physical conditions given in Table 4.9. The grid steps for calculations are

Table 4.9 Physcal parameters corresponding to [5]

$\hat{k}$	94300000	$\hat{\alpha}$	10^{-12}	$\hat{\beta}$	0.00133	ε	0.03
k_{ff}	3	$\bar{\theta}_0$	7.028	$t_0/l\rho$	2.62×10^{-7}	$cr_0/l\lambda e$	2.77×10^{-10}
R_0	0.0015	R	1	h_r	10^{-4}	τ	10^{-5}

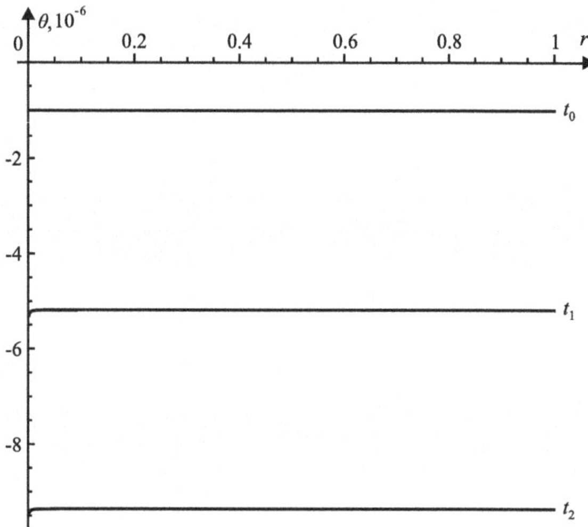

Fig. 4.12 Graphs of the temperature at different times for the parameters in Table 4.9

chosen as was described in Sect. 4.4.6. When modeling the dependence of the temperature on any parameter (see below), the other parameters are taken from Table 4.9. This set of parameters is known from the experiments described in [5]. The boundary condition for the heat equation on the lower base is the heat insulation:

$$\frac{\partial \theta}{\partial r}\bigg|_{r=R} = 0.$$

The modeling in this section is performed without taking the order function into account.

We note that the temperature distribution is the same on the cathode axis and on its lateral surface (Figs. 4.14 and 4.15).

There exists a strong nonlinear (and nonmonotone) dependence of the temperature on the voltage applied to the cathode (Ψ_2 in Fig. 4.4). The voltage influences the current density inside the cathode and the field strength which, in turn, influences the emission current and the cooling due to the Nottingham effect, see Fig. 4.16.

Since the formula for the emission current density 2.104 was obtained for a plane surface rather than for a conic vertex, it is necessary to introduce a correction coefficient, and the current density value is determined in calculations as follows:

$$j_{\text{eval}} = k_{\text{ff}} \cdot j_{\text{em}}, \tag{4.104}$$

where j_{eval} is the emission current density used in the modeling, k_{ff} is the correction coefficient ("form-factor"), and j_{em} is the computational current density obtained by formula (2.104). The linear variation in the emission current density due to the "form-factor" influences the temperature almost linearly, see Fig. 4.17. The value of this form-factor is the "adjusting" parameter of our model and determines the emission current density corresponding to that we know from the poor experimental data given in [5].

The cooling on the lower base is also significant, see Fig. 4.18. The boundary condition on the lower base is defined as [(see (4.14)]

$$\left. \frac{\partial \theta}{\partial \mathbf{n}} \right|_{r=R} = -\alpha_{\text{cool}}.$$

Finally, the radius of the top of the cathode rounding is also significant, see Fig. 4.19.

In the model, both the melting and the solidification are possible for the parameters close to the parameters in experiments described in [5], see Table 4.9. The graphs of the cathode cooling are given in Figs. 4.12, 4.13, 4.14 and 4.15. To model the heating above the melting temperature, we use the same initial condition as in Fig. 4.12 (time t_0) but decrease the form-factor. Figure 4.20 shows the graph of temperature for the form-factor $k_{\text{ff}} = 0.1$. The heating can also be obtained by decreasing the angle at the top of the cathode, see Fig. 4.21. Finally, we present the graph with increased rounding radius of the upper base (its influence on the thermal regime of the cathode was shown above) in Fig. 4.22. In this case, the initial temperature was the same as in Fig. 4.12. In this case, the influence of the Nottingham effect on the thermal regime is significantly less than that of the rounding radius R_0 in Table 4.9, and the cathode melts.

For not to large deviations of the model parameters from the experimental ones (those taken from [5]), we observe in our model either the absence of melting or the melting of the whole cathode.

For significant deviations of the model parameters from the experimental ones, the regime of local melting with subsequent solidification is possible (see above).

The dependence of the temperature on the dimensionless coefficient of thermal conductivity $\hat{k}$ is shown in Fig. 4.23. One can see that, for $\hat{k}$ of the order of 10^8, just as in the experiment, the dependence is practically linear. The temperature decreases with increasing $\hat{k}$.

Figure 4.24 presents the graph of the potential (the potential $\Psi_2 = 4\,\text{V}$ is applied to the cathode base, and it is assumed that the potential is zero at the top of the cathode because the distance between the anode and the top of the cathode is very small, see Sect. 4.2 for details).

In conclusion, we present the dependence of the temperature and the emission current density on the angle at the top of the cathode. The calculations were performed for the angles Θ varying from 5 to 27.5°, which corresponds to the total angle at

Fig. 4.13 Graph of the
temperature at time t_2 for the
parameters in Table 4.9,
large scale

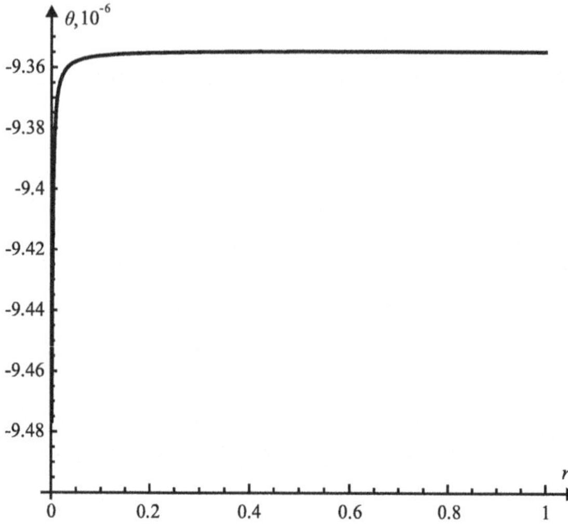

Fig. 4.14 Graph of the
temperature on the cathode
axis for the parameters in
Table 4.9

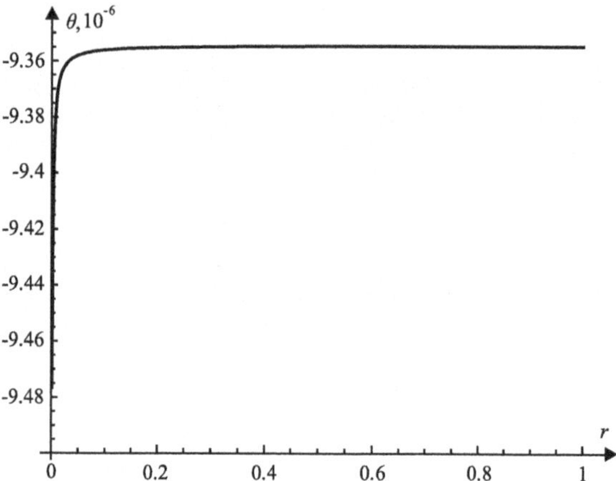

the vertex varying from 10 to 55°, while the cathode height and the upper base area
remain constant. The results are shown in Figs. 4.25 and 4.26. One can see that the
dependence of the temperature and especially of the emission current on the angle
is rather significant.

4.8   Formation of Melting and Crystallization Nuclei
in the Model

We assume that the domain is occupied by one (solid) phase, i.e., $\varphi = -1$. In this case,
for $\theta \neq \bar{\theta}_0$ ($\bar{\theta}_0$ is the dimensionless melting temperature), Eq. (4.13) has a nonzero

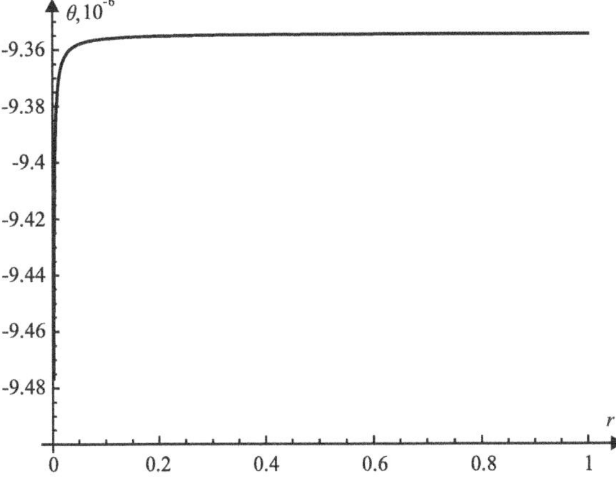

Fig. 4.15 Graphs of the temperature on the cathode surface for the parameters in Table 4.9

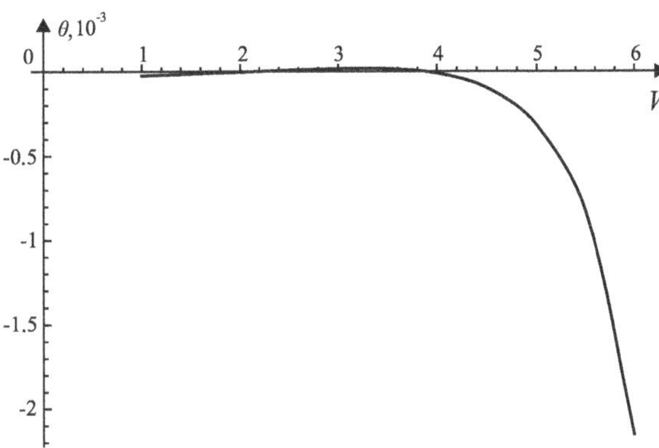

Fig. 4.16 Dependence of the relative temperature of the upper base on the voltage for the parameters in Table 4.9

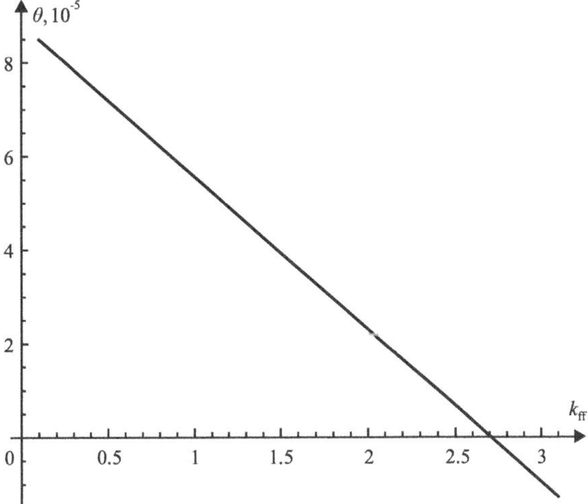

Fig. 4.17 Dependence of the relative temperature of the top of the cathode on the "form-factor" for the parameters in Table 4.9

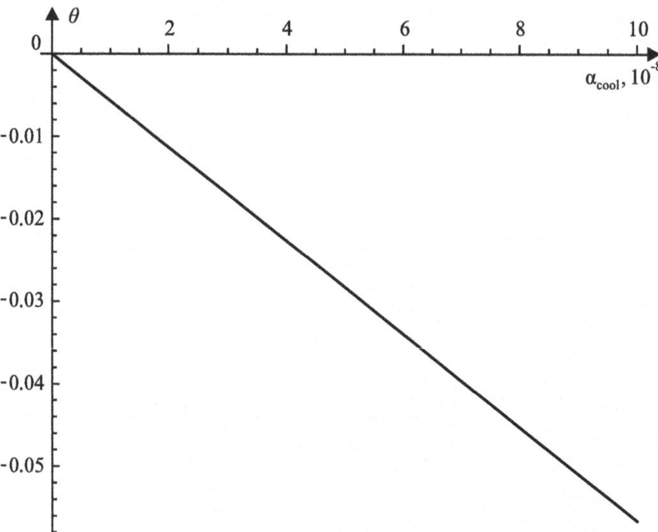

Fig. 4.18 Dependence of the relative temperature of the upper base on the coefficient in the right boundary condition determining the cooling of the lower base for the parameters in Table 4.9

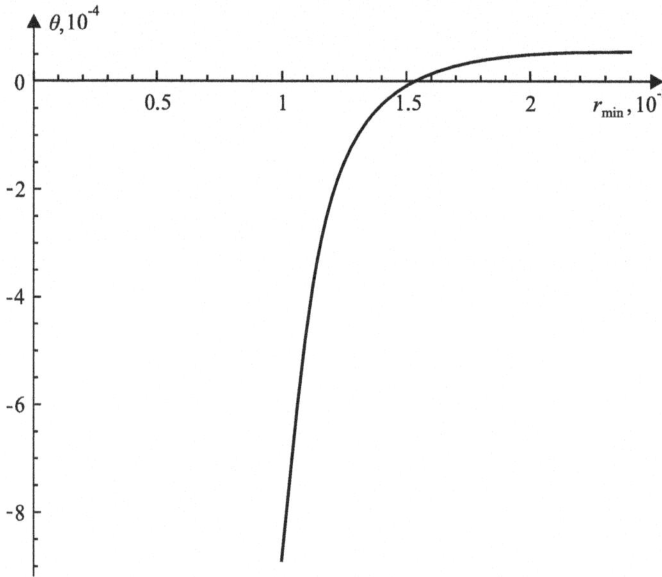

Fig. 4.19 Dependence of the relative temperature of the top of the cathode on the rounding radius for the parameters in Table 4.9

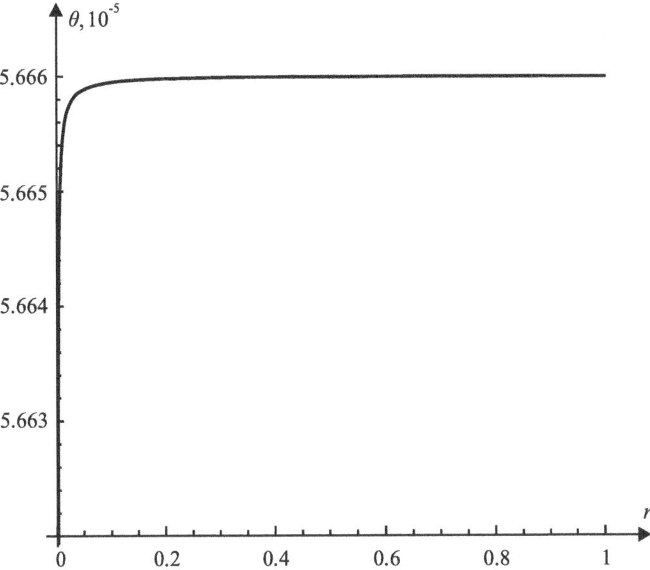

Fig. 4.20 Graph of the temperature for $k_{\text{ff}} = 0.1$ at the final time. The other parameters are taken from Table 4.9

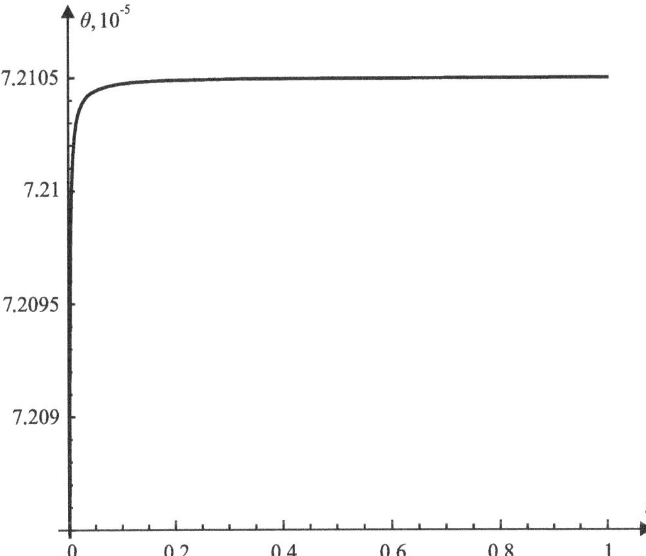

Fig. 4.21 Graph of the temperature for $\Theta = 5.7°$ at the final time. The other parameters are taken from Table 4.9

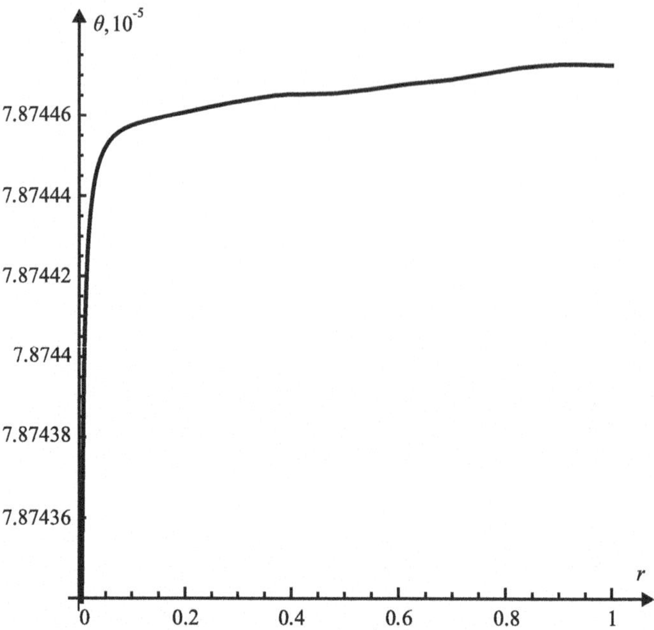

Fig. 4.22 Graph of the temperature for $R_0 = 0.005$ at the final time. The other parameters are taken from Table 4.9

Fig. 4.23 Dependence of the relative temperature of the upper base on $\hat{k}$ for the parameters in Table 4.9

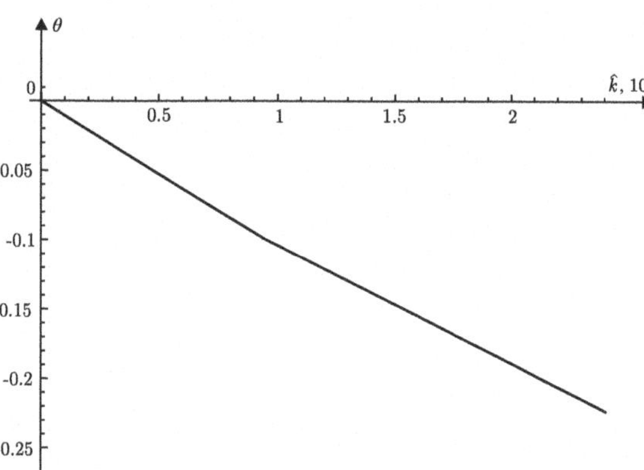

right-hand side which changes the order function by a quantity approximately equal to

$$\frac{1}{\hat{\alpha}\varepsilon} \int\limits_{t_0}^{t_0+t} (\theta - \bar{\theta}_0)\mathrm{d}t' \tag{4.105}$$

for small t. We note that the values $\varphi = \pm 1$ are stable, $\varphi(1 - \varphi^2) > 0$ for $0 < \varphi < 1$, and $\varphi(1 - \varphi^2) < 0$ for $-1 < \varphi < 0$.

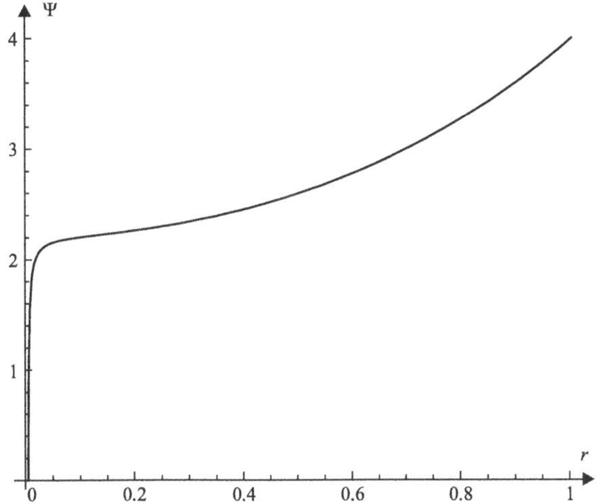

Fig. 4.24 Graphs of the potential $\Psi(r)$

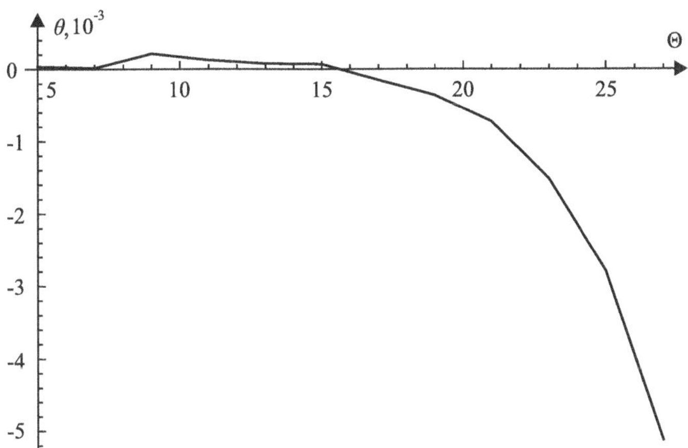

Fig. 4.25 Dependence of the temperature on the angle at the top of the cathode, from [3]

The contribution of nonlinearity to the variation of φ for small t is approximately equal to

$$\frac{1}{\hat{\alpha}\varepsilon^2} \int\limits_{t_0}^{t_0+t} \varphi(1 - \varphi^2)\mathrm{d}t'. \tag{4.106}$$

It is clear that the contribution of (4.106) due to the multiplier ε^{-2} is greater than the contribution of (4.105). Therefore, the nonlinearity also suppresses the influence of the temperature for "not too large" values of $\theta - \bar{\theta}_0$ even if the melting temperature

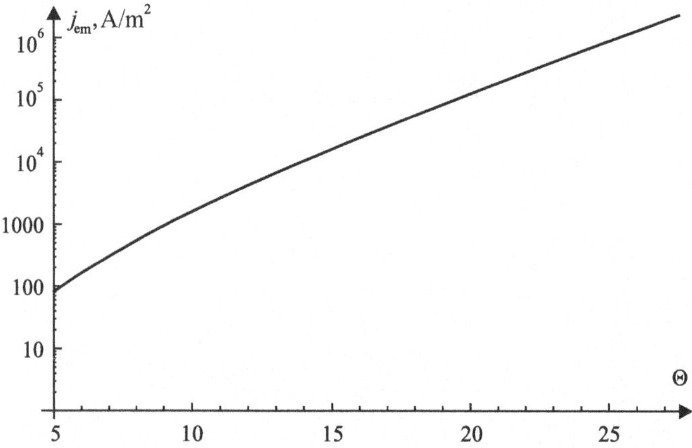

Fig. 4.26 Dependence of the emission current density on the angle at the top of the cathode, $\theta = -0.1$, from [3]. The ordinate axis is logarithmic

is exceeded, and the function φ is still approximately equal to -1, see Fig. 4.27. It is clear that such a behavior of the solution contradicts the meaning of the process under study.

On the other hand, it is known that the new phase formation starts from the appearance of nuclei, i.e., some inhomogeneities carrying the origination of this new phase inside them [8].

To resolve this contradiction with the physical meaning, which is a consequence of the chosen model, we propose our own method for introducing the nuclei which does not lie in the framework of the phase field system. This method differs from that proposed in [8], where it is assumed that a nucleus is a small volume of the new phase inside the volume of the old phase. We introduce a nucleus as an unstable state in a small volume of the old phase, i.e., one can say that we artificially create an unstable "mushy region" (in contrast to the stable one, see Sect. 3.4) in a small volume of the old phase. Of course, we meet the problem of conditions under which a "mushy region" can be created.

For the Allen–Cahn equation with the right-hand side, which describes the order function, it is known that, in the case of homogeneous equation, the states $\varphi = \pm 1$ are stable (this was already noted above) and the state $\varphi = 0$ is unstable. Therefore, it is natural to assume that the phase nucleus is a domain, where $\varphi \approx 0$. In this case, as is known from physical experiments, some nuclei lead to in the appearance of a new phase, and some of them disappear, and a nucleus of the liquid phase can disappear even at a temperature exceeding the melting temperature. In the case under study, the phase nuclei exhibit exactly such properties and the development of a nucleus depends on the relationship between its dimensions and the overheating value.

We shall determine the condition under which a nucleus can develop. In our case, the development means that, on the interval under study, φ takes a value different from the value at the appearance of the nucleus; namely, if, for example, the value

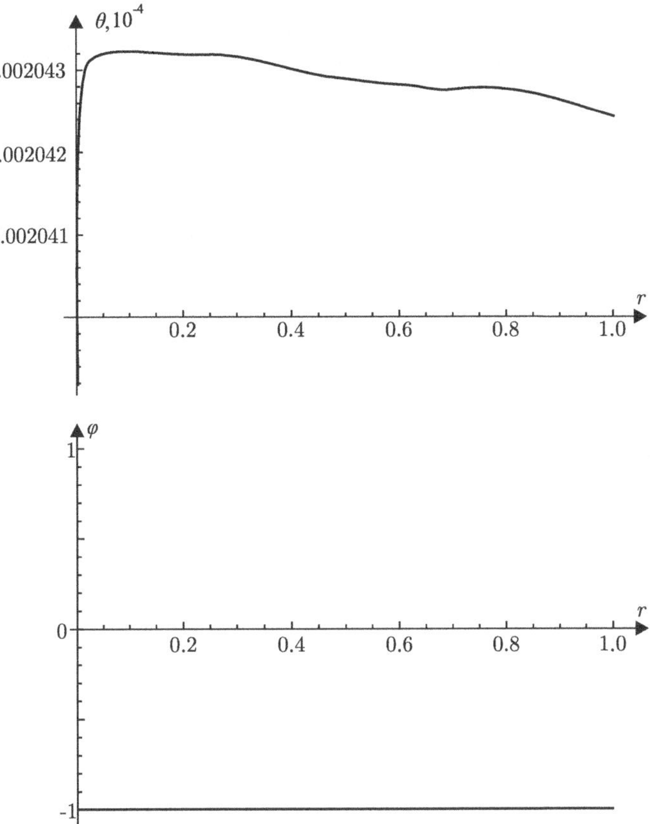

Fig. 4.27 Graphs of the temperature and order function after the melting temperature is exceeded in the case of calculations by the usual algorithm ; $0.0022 \le r \le 1.0013$, from [3]

$\varphi \approx 0$ arises from the value $\varphi = -1$, then the development of the nucleus means the evolution of φ to $+1$, and its disappearance means the evolution to -1. Therefore, the condition of the nucleus development can be formulated in terns of the time-derivative of the integral

$$\iiint_{V} \varphi(r,t)\mathrm{d}V, \tag{4.107}$$

where V is the cathode volume (a truncated cone). First, we note that

$$\iiint_{V} f(r,t)\mathrm{d}V = \int_{R_0}^{R} r^2 f(r,t)\mathrm{d}r \cdot \int_{0}^{\pi} \mathrm{d}\phi \cdot \int_{-\Theta}^{\Theta} |\sin \vartheta|\mathrm{d}\vartheta$$

$$= 2\pi(1 - \cos \Theta) \int_{R_0}^{R} r^2 f(r,t)\mathrm{d}r.$$

By R_0 and R we denote the left and right boundaries of the interval of integration. The derivative of (4.107) with respect to t can be determined from the balance relation. For this, we integrate the left- and right-hand sides of (4.13) over the cathode volume and obtain

$$
\int_{R_0}^{R} \varepsilon \hat{\alpha} r^2 \frac{\partial \varphi}{\partial t} dr - \int_{R_0}^{R} \varepsilon \hat{\beta}^2 \frac{\partial}{\partial r} r^2 \frac{\partial \varphi}{\partial r} dr
$$

$$
= \int_{R_0}^{R} \frac{1}{\varepsilon} r^2 (\varphi - \varphi^3) dr + \int_{R_0}^{R} \chi r^2 (\theta - \bar{\theta}_0) dr. \qquad (4.108)
$$

We assume that (and this is confirmed by numerical experiments) that the order function is less dependent on the polar angle and is completely independent of the azimuth angle. The same concerns the temperature.

Therefore, when integrating over the spherical layers between sections of the cone (cathode) by spheres centered at the cone vertex, we always obtain integrals over the radial variable multiplied by the same constant equal to $2\pi(1 - \cos \Theta)$. It is clear that this constant can be omitted in all terms.

Now we determine the class of functions used to choose the nucleus. As was noted in Sect. 3.4, the evolution of nonlinear parabolic waves can approximately be described by products of simple waves. For the Allen–Cahn equation, the simple waves have the form

$$
\hat{\varphi} = \tanh\left(\pm \frac{x - vt}{\varepsilon \hat{\beta} \sqrt{2}}\right),
$$

where the velocity v is determined by external conditions.

As the nuclei, we consider the family of functions of the form

$$
\varphi = \frac{A}{4}\left(1 + \tanh\left(\frac{r - (r_0 - \delta/2)}{\varepsilon \hat{\beta} \sqrt{2}}\right)\right)\left(1 - \tanh\left(\frac{r - (r_0 + \delta/2)}{\varepsilon \hat{\beta} \sqrt{2}}\right)\right) - 1, \quad (4.109)
$$

where r_0 is the position of the nucleus center and δ is its width, see Fig. 4.28. The constant A determines the nucleus amplitude.

For $A = 2$ and a sufficient value of δ, such a function is the simplest representative of the "kink gas" considered in [12]. For $A = 1$, the function (4.109) ranges in the strip $[-1, 0)$, and for a small excess of the amplitude A over 1, it has the form shown in Fig. 4.28. It is natural to take such a function as the nucleus of the liquid phase in the solid phase.

More precisely, the nucleus must satisfy the following conditions.

1. At a temperature greater than the melting temperature, the nucleus must in general increase, i.e., as the time increases, its maximum must attain a value close to $+1$ on a certain interval (from the standpoint of the properties of the order function, this means that there arises a melted layer).

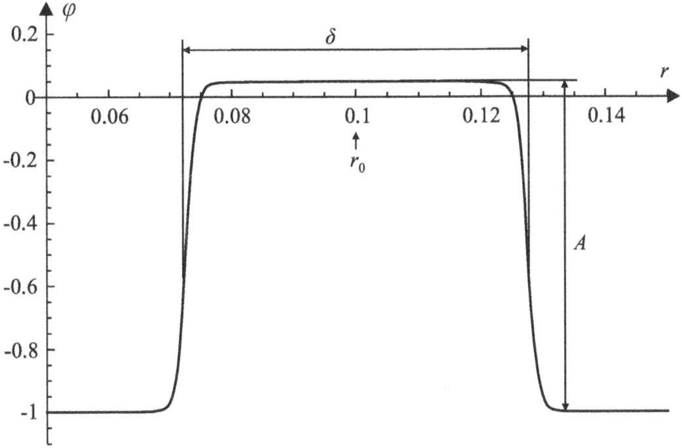

Fig. 4.28 Melting nucleus, from [3]

2. The rate of variation in the temperature must remain unchanged when the nucleus
   is introduced.

The nucleus is introduced by the following algorithm. We calculate the temperature till the time t_0 at which it begins to exceed the melting temperature. At this time, the computation stops (the modeling of the process is interrupted) and then starts with the calculated temperature taken as the initial condition for θ and with φ given by (4.109).

The equations of phase field system (4.12), (4.13) can be used to write the balance equation, i.e., an expression for

$$\frac{d}{dt} \int_{R_0}^{R} r^2 \theta dr \bigg|_{t=t_0}.$$

We integrate the heat equation (4.12) and obtain

$$\frac{d}{dt} \int_{R_0}^{R} r^2 (\theta - \bar{\theta}_0) dr = \hat{k} R^2 \frac{\partial \theta}{\partial r}\bigg|_{r=R} - \hat{k} R_0^2 \frac{\partial \theta}{\partial r}\bigg|_{r=R_0}$$

$$- \frac{1}{2} \int_{R_0}^{R} r^2 \frac{\partial \varphi}{\partial t} dr + \int_{R_0}^{R} r^2 \hat{F} dr. \tag{4.110}$$

Substituting the expression $\partial \varphi / \partial t$ from (4.108) in this relation and taking into account that $\partial \varphi / \partial r = 0$ for $r = R_0, R$, we obtain

$$\frac{\mathrm{d}}{\mathrm{d}t} \int_{R_0}^{R} r^2(\theta - \bar{\theta}_0)\mathrm{d}r = \hat{k}R^2 \left.\frac{\partial\theta}{\partial r}\right|_{r=R} - \hat{k}R_0^2 \left.\frac{\partial\theta}{\partial r}\right|_{r=R_0} - \frac{1}{2\varepsilon^2} \int_{R_0}^{R} r^2(\varphi - \varphi^3)\mathrm{d}r$$

$$- \frac{1}{2\varepsilon} \int_{R_0}^{R} \chi r^2(\theta - \bar{\theta}_0)\mathrm{d}r + \int_{R_0}^{R} r^2 \hat{F}\mathrm{d}r. \qquad (4.111)$$

Here we took into account that the equation for the order function φ is solved by the iteration method, see Sect. 4.4.2. This means that the equations for the order function are solved until the derivative $\partial\varphi/\partial t$ becomes sufficiently small. It follows from (4.111) that all terms except for the third are continuous for $t = t_0$, and the third term experiences a jump equal to the integral

$$\frac{1}{\varepsilon} \int_{R_0}^{R} r^2(\varphi - \varphi^3)\mathrm{d}r, \qquad (4.112)$$

where φ is a function of the form (4.109). For continuity, this term must be equated to zero. Figure 4.29 illustrates the behavior of this term, which is denoted by I, depending on the parameters determining it.

For $t > t_0$, the sum of the third and fourth terms in the right-hand side of (4.111) is small, i.e.,

$$\frac{d}{dt} \int_{R_0}^{R} r^2 \varphi \mathrm{d}r \bigg|_{t=t_0}.$$

The dependence of this expression on the number of iterations at a time step is shown in Table 4.10.

Thus, the main contribution to the right-hand side of Eq. (4.111) is made by the total capacity of the heat sources, and the nucleus develops at a positive temperature.

We determine the nucleus parameters r_0, δ, and A starting from the condition that (4.112) is equal to zero. The parameter A can be determined for given r_0 and δ. Further, it is reasonable to place the nucleus at the point of maximum temperature (on the cathode axis), i.e., r_0 is determined by the condition

$$\theta(r_0, 0) = \max_{R_0 \leq r \leq R} \theta(r, 0)$$

at a time step after which the nucleus is added. Finally, it is necessary to determine δ. We consider the graph of the dependence $A(\delta)$ for a fixed r_0, see Fig. 4.30. One can see that it resembles a hyperbolic dependence, and as the nucleus width it is reasonable to take a width δ for which the amplitude A varies weakly. In this case, the graph in Fig. 4.30 is almost independent of the choice of r_0, and as the width, one can take, for example, the value $\delta = 1500\,\varepsilon\hat{\beta}$.

Fig. 4.29 Dependence of I on the nucleus position r_0, the width δ, and the amplitude A. The nucleus width δ is given by the formula $\delta = c_\delta \varepsilon \hat{\beta}$, from [3]

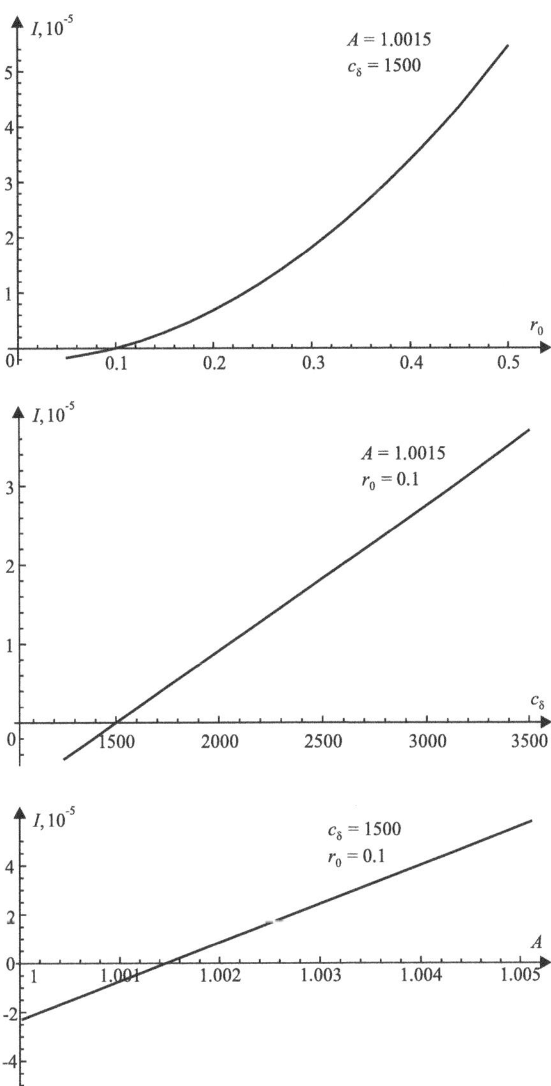

Here we present the results of modeling of the melting process, where we introduce the liquid phase nucleus under the following assumptions.

The angle at the top of the cathode is decreased in order to decrease the influence of the Nottingham effect. In this section, this angle is taken to be equal to 11.4°. The same effect can also be obtained in the case of constant geometry by decreasing the form-factor in the Nottingham condition which, in turn, can be explained by the influence of effects typical of semiconductors (in particular, the structure defects or the influence of impurities) or by the influence of the problem geometry.

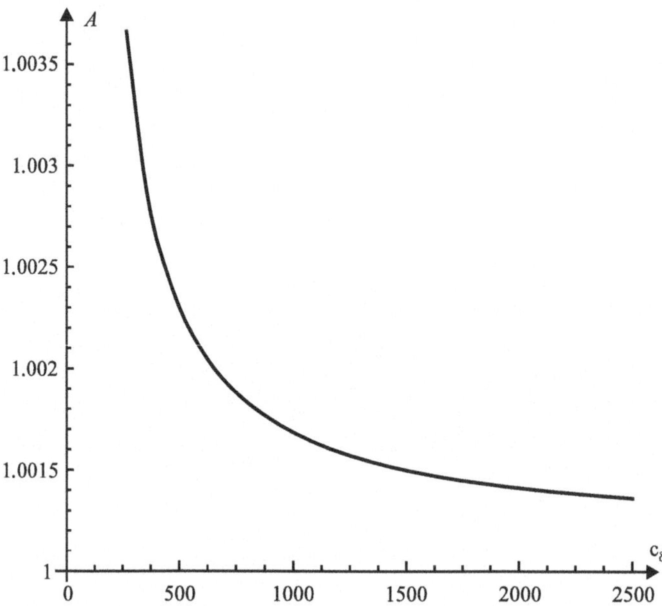

Fig. 4.30 Dependence of the amplitude A on the nucleus width δ for $r_0 = 0.1$. The nucleus width δ is given by the formula $\delta = c_\delta \varepsilon \hat{\beta}$, from [3]

Moreover, in the semiconductors, in contrast to metals, the specific conductivity increases with the temperature. Initially we assumed that this is possible in the transition to the liquid state when additional electrons appear in semiconductors due to the break of covalent bonds between atoms. Unbelievable, but later we found many experiments (see [6, Chap. 3, Sect. 1.2] and references therein) confirming that the conductivity of semiconductors has a jump at the melting point. Namely, the conductivity of silicon (near the melting point) in solid phase is equal to 5.8×10^4 $(\Omega\,m)^{-1}$, but in liquid phase, it is equal to 12×10^5 $(\Omega\,m)^{-1}$, see [6]. Thus, we have a jump of conductivity of silicon, and the conductivity after melting increases nearly by 20 times. Figure 4.31 shows an approximate curve of silicon conductivity constructed from experimental data collected from ten papers and books cited in [6]. Note that the temperature range interesting for us is $T_0 \pm 100\,K$, where T_0 is the melting point temperature (this follows from the results of numerical simulation, see Sect. 4.7).

Due to this, in the case of local melting, the conductivity increases jumpwise, which can result in a decrease in the Joule heat. But the exact value of variation in the conductivity has not yet been obtained. On the other hand, by general considerations, it is natural to assume that a part of free electrons from the melting region can increase the emission current density, which implies an increase in the Nottingham effect, i.e., in the energy carried away by the emitting electrons. In our model, we use a formula describing the Nottingham effect for the emission of electrons which does not take into account the above-cited increase in the flow of emitting electrons. Therefore, we

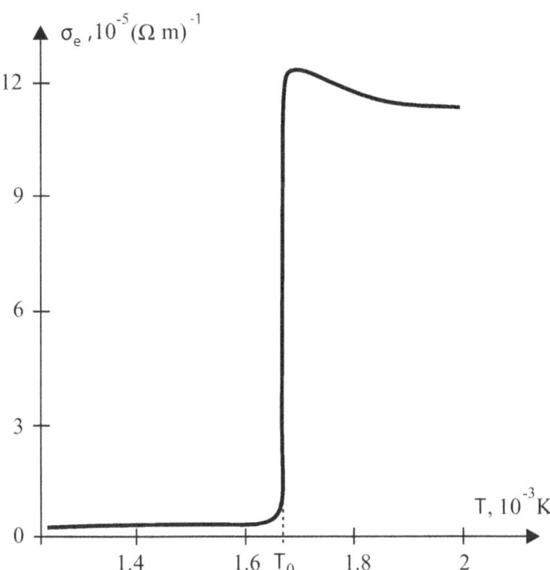

Fig. 4.31 Jump of silicon conductivity, T_0 is the melting point temperature

introduce an artificial "form-factor" in it, i.e., a correction coefficient which allows us to take into account both (hypothetical) phenomena described above, because both of them result in the cooling of the cathode material. In contrast to the preceding section, this form-factor is a function of time (temperature) and is determined by a curve of the form shown in Fig. 4.32. This determination of the form-factor permits taking into account the possible jumpwise increase in the number of electrons entering the top of the cathode when the melting region is being formed. The jump in the Nottingham condition results in that the cathode begins to cool after melting, which further results in its solidification and disappearance of the region occupied by the liquid phase, see below.

Now we illustrate the above considerations. In Fig. 4.33, we show a typical graph of the temperature above the melting temperature. In fact, the nucleus can be introduced at any time at which the temperature exceeds the melting temperature. We choose the temperature after several computation steps so that, with regard to the error, the calculated temperature is above the melting temperature. At such a temperature, we introduce a nucleus, see Fig. 4.33. The temperature in our algorithm is calculated from a given order function, and therefore, at the first moment after the nucleus introduction, the temperature remains almost unchanged by the second condition of the nucleus formation, but its local structure is changed (Fig. 4.34). Moreover, the parabola-shaped fragment of the temperature graph between the phase boundaries is typical of problems with interacting fronts [4]. In that paper, such a behavior of the temperature was obtained analytically. In Figs. 4.35 and 4.36, this parabolic fragment is "blurred" due to the action of the Nottingham boundary condition.

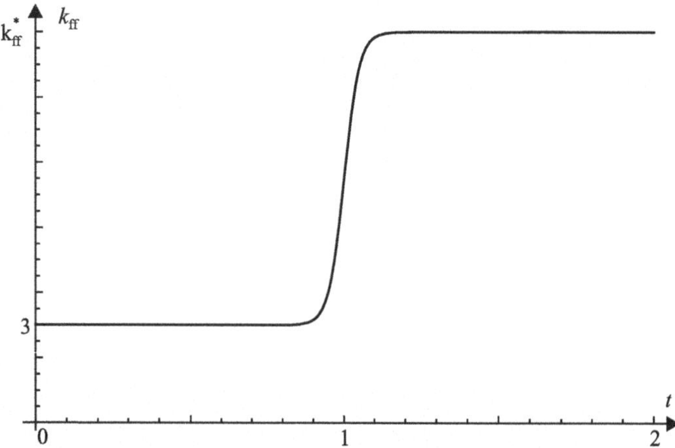

Fig. 4.32 Dependence $k_{\mathrm{ff}}(t)$, $k^*_{ff} = 12$

It follows from the balance relations that the nucleus develops as is shown in Fig. 4.35. Finally, the nucleus becomes a part of the liquid phase, see Fig. 4.36. The process of the nucleus development is accompanied by the heat release after the time of the nucleus formation (described by the term $-(\partial\varphi/\partial t)/2$ in the right-hand side of the heat equation) which is required for the phase transition of the cathode material, and hence the temperature decreases during the nucleus development and can even be lower than the melting temperature, as one can see in our figures.

There are two versions of the further modeling. In the first case, we start from the above considerations that the Nottingham effect must increase after formation of the liquid phase, which can result in the further cooling and collapse of the liquid phase region. For example, we considered the form-factor equal to $k^*_{\mathrm{ff}} = 12$. The graphs illustrating the further behavior of the process are shown in Fig. 4.37. One can see that the temperature actually continues to decrease, and the width of the liquid phase region decreases a little.

The relative temperature at which the free boundary stops and the nucleus collapses is related to the position of the free boundary, see (4.11). One can see that if the temperature maximum is at a farther distance from the top of the cathode, which is possible in the case of strong cooling on the lower base of the cathode and for a not very strong influence of the Nottingham effect, then the free boundaries stop and the nucleus disappears at a temperature that is closer to the melting temperature than that in the case, where the temperature maximum is located closer to the top of the cathode, just as in the figure in this section (i.e., for a greater curvature of the interface between phases).

Figure 4.39 shows that the free boundary moves very slowly (practically does not move) as compared with the decrease in the temperature, and hence, by condition (4.11), after the temperature becomes lower that the temperature obtained from this condition (i.e., lower than $\theta - \bar{\theta}_0 \approx -0.038$), the liquid phase region begins to

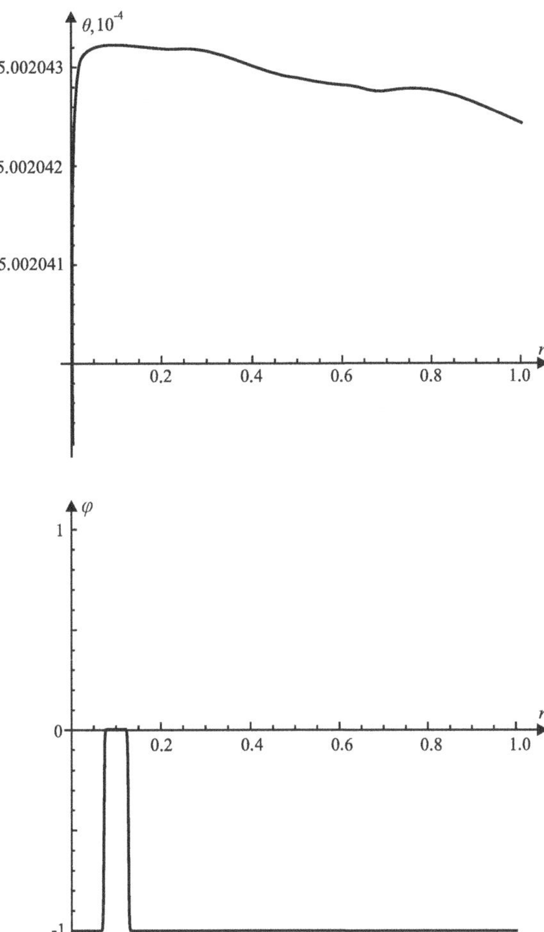

Fig. 4.33 Time t_0 at which the nucleus is introduced

narrow and finally decays. This process is similar to that described (for "incorrect" parameters) in Sect. 4.7.1, see Fig. 4.10. The slow motion of the free boundary can be explained by the small coefficient $\hat{\beta}^2 \sim 10^{-6}$ in front of the term with the Laplace operator in the Allen–Cahn equation. Figure 4.39 also shows that the temperature first falls rapidly because of the nucleus growth, and the fall becomes slower but still continues due to the increase in the form-factor. In this case, by the conditions under which the nucleus is added, the temperature remains almost unchanged at the initial time after the nucleus addition.

In the second version of modeling, we do not change the form-factor, and respectively, after formation of the liquid phase region, the phase transition requires a negligible amount of heat because of the slow motion of the free boundaries, and the temperature begins to increase (see Fig. 4.38). Although the temperature shown in the graph is lower than the melting temperature, it still continues to grow and

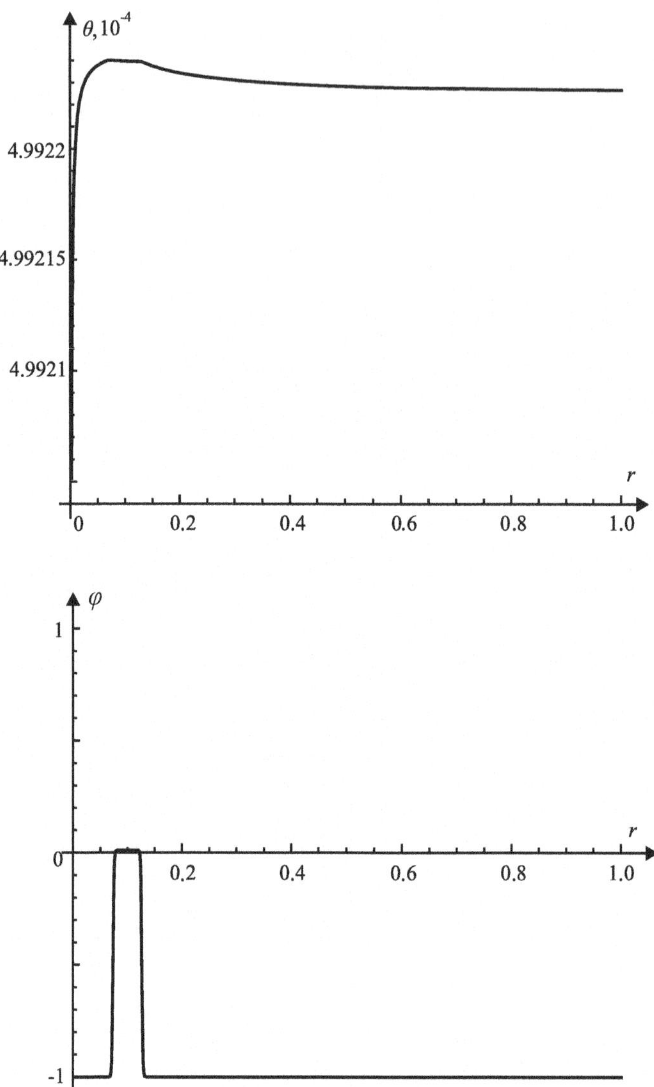

Fig. 4.34 Time t_1 the next after the nucleus addition

again becomes positive in the further modeling. In this case, the liquid phase region becomes broader and, in the course of time, the whole cathode will be melted starting from the top.

To conclude this section, we note that the algorithm for introducing the nucleus is here similar to that of artificial introduction of the nucleus in Sect. 4.7 for the parameters that are in no way related to experimental data.

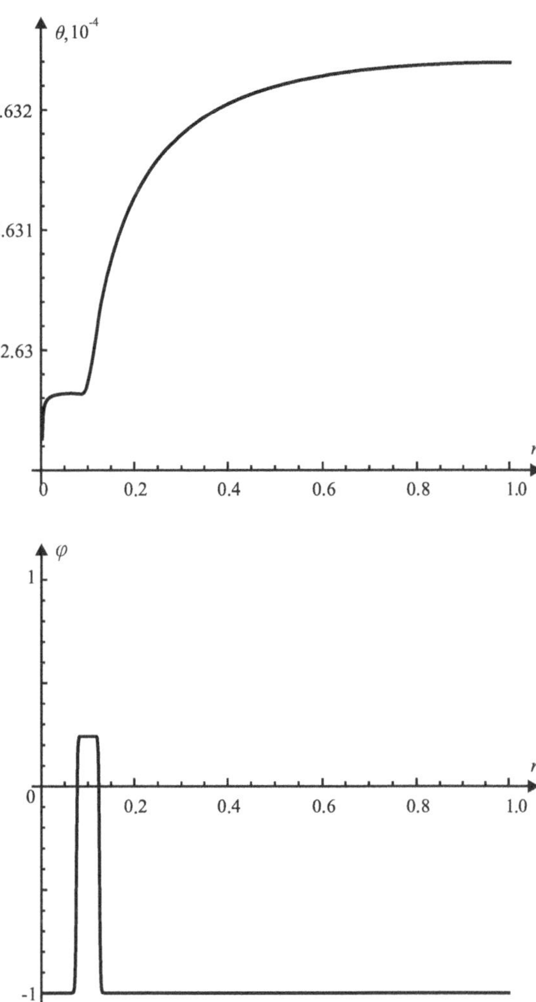

Fig. 4.35 Time t_2, the nucleus continues to grow

4.9 Conclusion

In this book, we proposed a system of equations describing the heat propagation in a small(nano)-dimensional cathode. This system was reduced to the dimensionless form with regard to the real physical parameters of the problem. It turned out that (due to small dimensions and a relatively high conductivity of silicon) the temperature rapidly levels in the cathode body except for the top (emission region), where intensive cooling occurs due to the Nottingham effect. It is shown that if the real parameter vary, then the cathode can either melt or remain solid, which confirms the adequacy of the model.

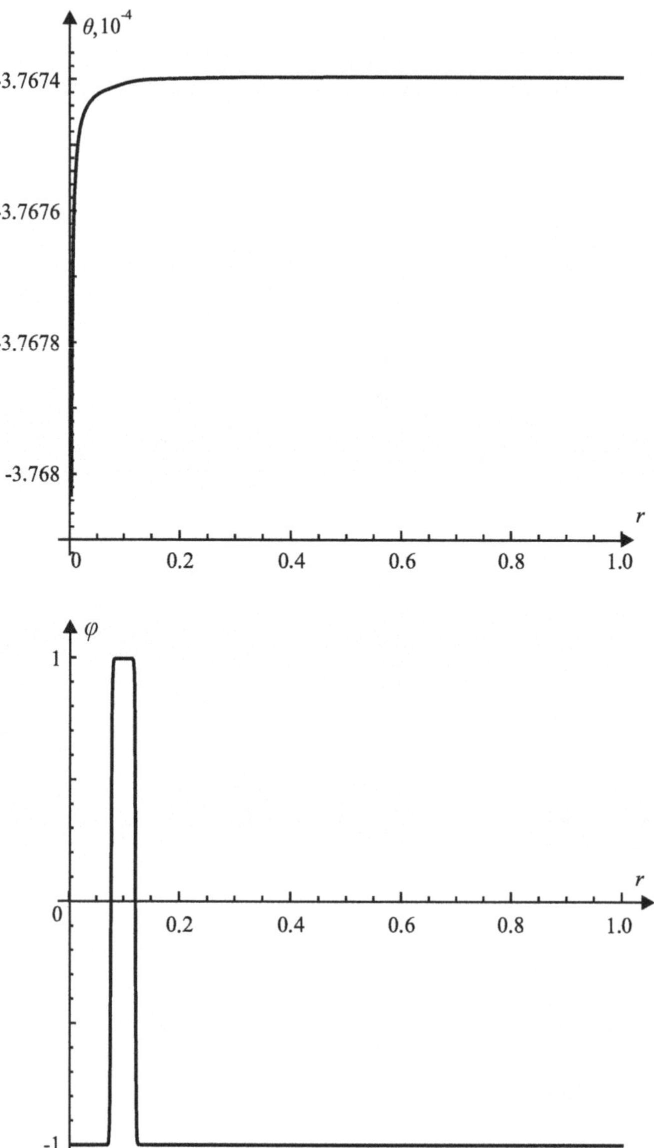

Fig. 4.36 Time t_3, the liquid phase region has been formed completely

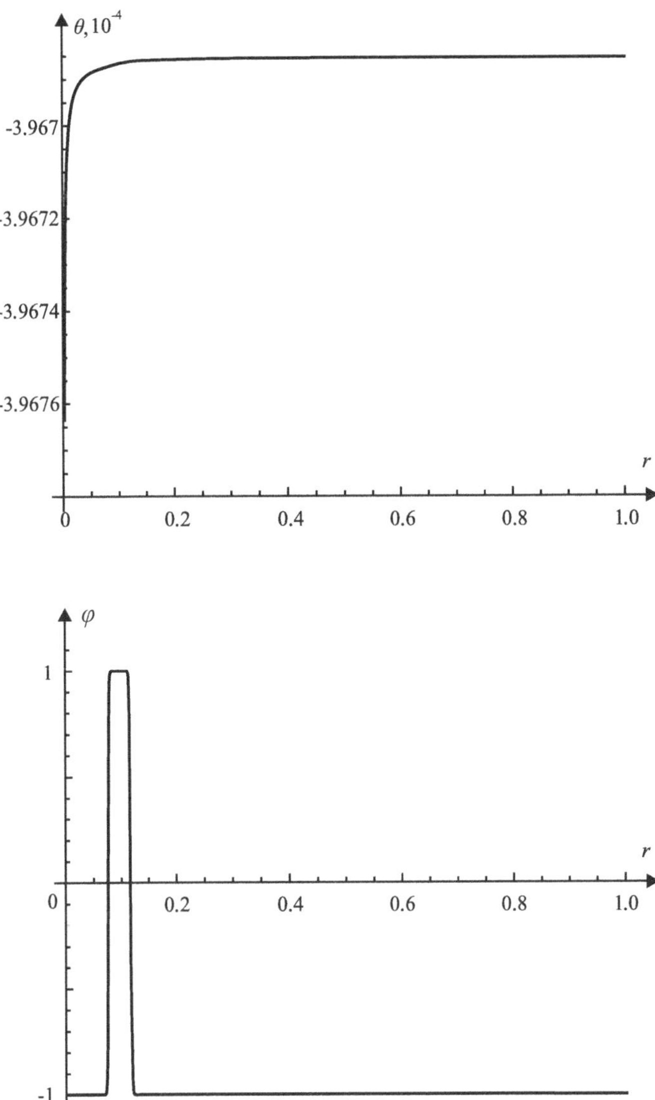

Fig. 4.37 Time t_4, the form-factor increases till $k_{\text{ff}}^* = 12$, the temperature decreases, and the liquid phase region stops to develop

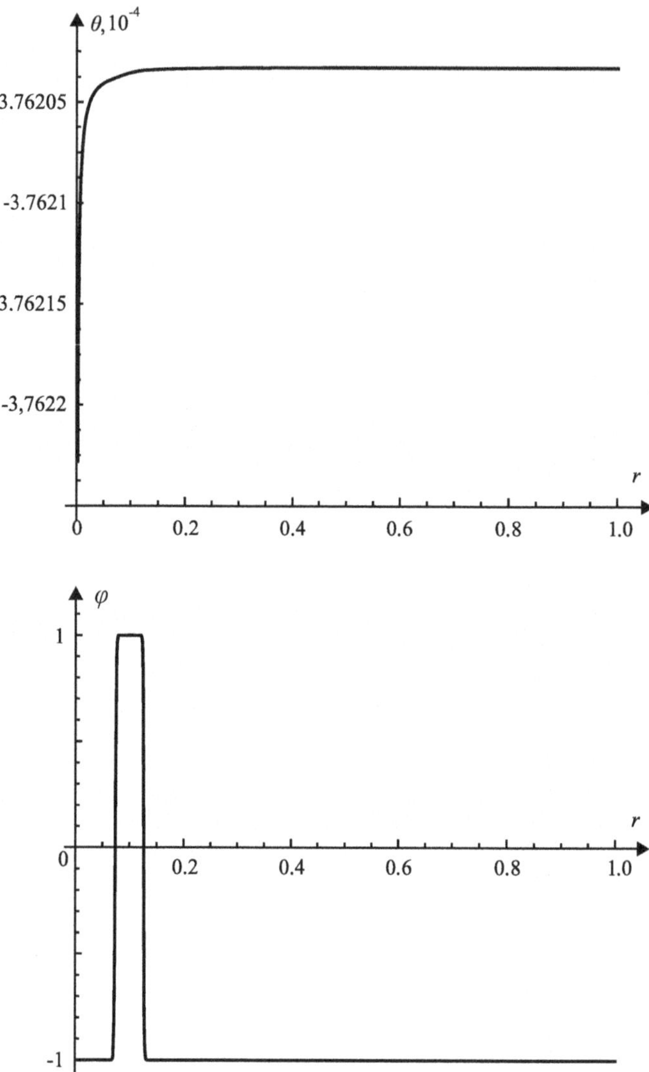

Fig. 4.38 Time t_4, the form-factor remains unchanged ($k_{\text{ff}}^* = 3$), the temperature begins to grow, the liquid phase region continues to develop

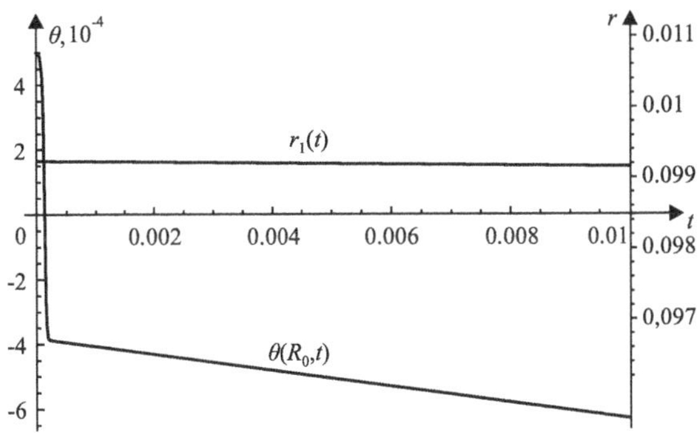

Fig. 4.39 Time-dependence of the cathode upper base temperature and the free boundary position the nearest to the upper base, from [3]

Table 4.10 Dependence of max φ and $\partial\varphi/\partial t$ on the nucleus parameters and the number of iterations

No.	A	r_0	c_δ	Number of iterations	max φ	$\partial\varphi/\partial t$
1	1.00714	0.1	300	10	0.02184	-8.309×10^{-9}
2	1.0015	0.1	1500	10	0.0046054	-8.4971×10^{-9}
3	1.00086	0.1	3000	10	0.0026491	-1.0844×10^{-8}
4	1.00712	0.4	300	10	0.021778	-1.3234×10^{-7}
5	1.00142	0.4	1500	10	0.0043599	-1.2492×10^{-7}
6	1.00072	0.4	3000	10	0.0022198	-1.3097×10^{-7}
7	1.00714	0.1	300	100	1	6.1171×10^{-11}
8	1.0015	0.1	1500	100	0.99995	-1.7708×10^{-9}
9	1.00086	0.1	3000	100	0.99985	-1.6791×10^{-9}
10	1.00712	0.4	300	100	1	9.2785×10^{-10}
11	1.00142	0.4	1500	100	0.99994	2.7561×10^{-8}
12	1.00072	0.4	3000	100	0.99977	-4.696×10^{-8}

In the book, we discovered the quantities that significantly influence the character of the process, i.e., the voltage applied to the cathode, the dimensionless coefficient $\hat{k}$, the rounding radius of the top of the cathode, the cooling on the lower base, the "form-factor" of the top of the cathode, and the quantities whose influence is not so important, i.e., the angle at the top of the cathode, and the cathode conductivity. This classification may be important in subsequent experimental works.

Passing to dimensionless variables shows that, because of small dimensions of the cathode, it is necessary to take into account the curvature of the interface between phases (in the case of appearance of the liquid phase, the melting). In macrosystems, this quantity is neglected as a rule, and the classical Stefan conditions are considered on the interface between phases. But here the curvature determines the position of the interface, namely, the temperature on the boundary is proportional to the curvature.

In general, in the case of a small radius of the upper base, the free boundary cannot "reach" the top of the cathode (emission region), where the cooling occurs due to the Nottingham effect. But under the conditions realizing the poor experimental data available to us, the melting region does not at all arise due to the Nottingham effect and the temperature levelling.

There is another obvious reserve to suppress the substrate melting or cooling as this is usually done in microelectronics.

Here it is interesting to mention the competition between the cooling and melting mechanisms, see below. A decrease in the diameter of the lower base decreases the cooling through the substrate but increases the stability of the melting region in the middle.

We also recall some preliminary considerations which show that the problem of modeling of the field emission from a nanocathode must be solved with regard to the cathode geometry that plays a significant role in the heat balance. The transition to dimensionless quantities shows that condition (4.11) is satisfied on the free boundaries in the case of pointed conic cathode. This means that the upper free boundary remains in the region, where the cathode temperature deviation from the melting temperature is negative (i.e., in the regions where the cathode is overcooled). This means that the geometry "helps" the melting, because the melting region moves into the region where the temperature is less than the melting temperature.

In Fig. 4.40, the transition from -1 to $+1$ for $r = r_1(t)$ is described by the formula

$$\varphi \approx \omega\left(\frac{r - r_1(t)}{\varepsilon}\right),$$

where the function ω was introduced above. This transition is related to the condition

$$\left.(\bar{\theta} - \bar{\theta}_0)\right|_{r=r_1(t)} = -\hat{\beta}\frac{2}{r_1(t)}, \tag{4.113}$$

which was mentioned above. The transition from $+1$ to -1 for $r = r_2(t)$ is related to the following expression for φ:

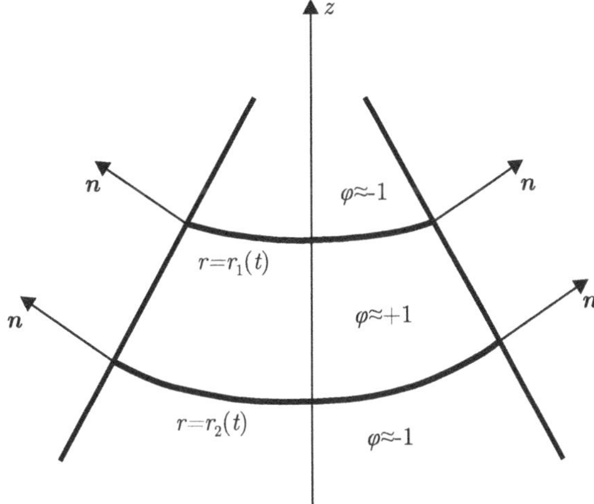

Fig. 4.40 Shape of a cathode with free boundaries of negative curvature

$$\varphi \approx \omega\left(\frac{r_2(t)-r}{\varepsilon}\right),$$

with the boundary condition

$$\left.\left(\bar{\theta}-\bar{\theta}_0\right)\right|_{r=r_2(t)} = \hat{\beta}\frac{2}{r_2(t)}. \tag{4.114}$$

In contrast to condition (4.113), condition (4.114) means that the melting region stops to expand at a temperature above the melting temperature, i.e., the Gibbs–Thomson condition hinders the melting.

Taking into account the cooling of the top of the cathode due to the Nottingham effect which prevents the free boundary from motion to the top, we can say that the Nature helps the cathode to stay alive.

Of course, the results presented above are far from being final. For more detailed investigations, it is necessary to consider the mechanical stresses arising in the cathode and to determine its optimal shape (this can really be done if we restrict our study, for example, to bodies of revolution).

References

1. Chapman, B., Jost, G., van der Pas, R.: Using OpenMP: Portable Shared Memory Parallel Programming. The MIT Press (2008)
2. Cook, S.: Cuda Programming: A Developer's Guide to Parallel Computing with GPUs. Elsevier (2013)

3. Danilov, V.G., Gaydukov, R.K., Kretov, V.I., Rudnev, V.Y.: Modelling of liquid nuclei generation for field-emission silicon nanocathode. IEEE Trans. Electron Devices **61**(12), 4232–4239 (2014)
4. Danilov, V.G., Omel'yanov, G.A., Radkevich, E.V.: Hugoniot-type conditions and weak solutions to the phase-field system. Eur. J. Appl. Math. **10**, 55–77 (1999)
5. Dyzhev, N.A., Gudkova, S.A., Makhiboroda, M.A., Fedirko V, A.: Investigation of emussion properties of silicon cathodes of different geometry. In: Bykov, D.V. (Ed.) Vacuum Science and Technics, Material of XII Scientific-Technical Conference with Participation of Foreign Specialists, pp. 221–224. MIEM, Moscow (2005). (in Russian)
6. Glazov, V.M., Chizhevskaia, S.N., Glagoleva, N.N.: Liquid Semiconductors. Springer (1969)
7. Iyengar, S., Jain, R.: Numerical Methods. New Age International Ltd (2009)
8. Kolmogorov, A.N.: On the statistical theory of metal crystallization. Izvestiya Akademii Nauk SSSR. Ser. Matematicheskaya **1**(3), 355–359 (1937). (in Russian)
9. Omel'yanov, G.A., Rudnev, V.Y.: Interaction of free boundaries in the modified Stefan problem. Nonlinear Phenom. Complex Syst. **7**(3), 227–237 (2004)
10. Rauber, T., Rünger, G.: Parallel Programming For Multicore and Cluster System. Springer (2010)
11. Samarskii, A.A.: The Theory of Difference Schemes. Marcel Dekker Inc. (2001)
12. Ward, M.J., Reyna, L.G.: Resolving weak internal layer interactions for the ginzburg-landau equation. Eur. J. Appl. Math. **5**, 495–523 (1994)

The manufacturer's authorised representative in the EU is Springer
Nature Customer Service Centre GmbH, Europaplatz 3, 69115 Heidelberg,
Germany. If you have any concerns regarding our products, please
contact ProductSafety@springernature.com

Printed and bound by CPI Group (UK) Ltd, Croydon, CR0 4YY
28/11/2025
02007706-0007